T0251605

The Role of the Outdoors in Residential Environments for Aging

The Role of the Outdoors in Residential Environments for Aging has been co-published simultaneously as *Journal of Housing for the Elderly*, Volume 19, Numbers 3/4 2005.

The Role of the Outdoors in Residential Environments for Aging

Susan Rodiek, PhD, NCARB
Benyamin Schwarz, PhD
Editors

The Role of the Outdoors in Residential Environments for Aging has been co-published simultaneously as *Journal of Housing for the Elderly*, Volume 19, Numbers 3/4 2005.

Routledge
Taylor & Francis Group
NEW YORK AND LONDON

First published by
The Haworth Press, Inc.
10 Alice Street
Binghamton, N Y 13904-1580

This edition published 2011 by Routledge

Routledge
Taylor & Francis Group
711 Third Avenue
New York, NY 10017

Routledge
Taylor & Francis Group
2 Park Square, Milton Park
Abingdon, Oxon OX14 4RN

The Role of the Outdoors in Residential Environments for Aging has been co-published simultaneously as *Journal of Housing for the Elderly*™, Volume 19, Numbers 3/4 2005.

© 2005 by The Haworth Press, Inc. All rights reserved. No part of this work may be reproduced or utilized in any form or by any means, electronic or mechanical, including photocopying, microfilm and recording, or by any information storage and retrieval system, without permission in writing from the publisher.

The development, preparation, and publication of this work has been undertaken with great care. However, the publisher, employees, editors, and agents of The Haworth Press and all imprints of The Haworth Press, Inc., including The Haworth Medical Press® and Pharmaceutical Products Press®, are not responsible for any errors contained herein or for consequences that may ensue from use of materials or information contained in this work. With regard to case studies, identities and circumstances of individuals discussed herein have been changed to protect confidentiality. Any resemblance to actual persons, living or dead, is entirely coincidental.

The Haworth Press is committed to the dissemination of ideas and information according to the highest standards of intellectual freedom and th free exchange of ideas. Statements made and opinions expressed in this publication do not necessarily reflect the views of the Publisher, Directors, management, or staff of The Haworth Press, Inc., or an endorsement by them.

Cover design by Deborah Huelsbergen
Cover photo "Woman at Window," © 2003, Susan Rodiek.
Other photos by Benyamin Schwarz.

Library of Congress Catalog-in-Publication Data

The role of the outdoors in residential environments for aging / Susan Rodiek, Benyamin Schwarz, editors.
        p. cm.
        "The Role of the Outdoors in Residential Environments for Aging has been co-published simultaneously as Journal of Housing for the Elderly, Volume 19, Numbers 3/4 2005."
        Includes bibliographical references and index.
        ISBN-13: 978-0-7890-3243-0 (hard cover : alk. paper)
        ISBN-10: 0-7890-3243-0 (hard cover : alk. paper)
        ISBN-13: 978-0-7890-3244-7 (soft cover : alk. paper)
        ISBN-10: 0-7890-3244-9 (soft cover : alk. paper)
        1. Landscape architecture for older people. 2. Older people–Dwellings–Psychological aspects. 3. Aging–Environmental aspects. 4. Environmental psychology. I. Rodiek, Susan. II. Schwarz, Benyamin.
SB475.9.A35R65 2006
712'.708696-dc22

                                                                              2005034366

The Role of the Outdoors in Residential Environments for Aging

CONTENTS

ABOUT THE EDITORS

Susan Rodiek, PhD, NCARB, teaches architectural design at Texas A&M University, where she is Associate Director of the Center for Health Systems & Design. Her work applies professional practice and environment-behavior expertise to facilities for aging, healthcare, and therapeutic garden environments.

Her recent research has focused on how outdoor access may benefit both residents and staff in long term care facilities for aging. Currently Dr. Rodiek is developing a multimedia educational series to translate evidence-based design principles into practical solutions to improve the quality of outdoor space for older adults. Funded by SBIR grant 1 R43 AG 024786-01 from the National Institute on Aging (NIA), this award-winning educational tool targets a broad range of industry professionals, including architects, landscape architects, long term care providers, policy planners, and consumer advocates.

Dr. Rodick is Coordinator of the combined Environment-Gerontology Network of the Environmental Design Research Association (EDRA), and the International Association of Person-Environment Studies (IAPS). She is an Advisory member of the Texas Healthy Aging Network and the Program on Health Promotion and Aging, School of Rural Public Health of the Texas A&M University System Health Sciences Center. Dr. Rodiek holds a doctorate in Architecture from Cardiff University, and is a nationally-registered architect with more than twenty years' experience in both landscape and architectural projects. She is on the editorial board of the *Journal of Therapeutic Horticulture*, and publishes in the areas of environmental gerontology, healthcare and therapeutic garden environments.

Benyamin Schwarz, PhD, is Professor in the Department of Architectural Studies at the University of Missouri-Columbia. His teaching specialty areas include design fundamentals, environmental analysis, housing concepts and issues, design studio, architectural programming, and environmental design for aging. He received his bachelor's degree in Architecture and Urban Planning from the Technion, the Institute of

Technology of Israel, and his PhD in Architecture from The University of Michigan with an emphasis on Environmental Gerontology. He designed numerous facilities for the elderly in Israel and in the U.S. His research addresses issues of long-term care settings in the United States and abroad, environmental attributes of dementia special care units, assisted living arrangements, and international housing concepts and issues. Dr. Schwarz has been the editor of the *Journal of Housing for the Elderly* since 2000. He is the author of *Nursing Home Design: Consequences of Employing the Medical Model*. His co-edited books with Leon A. Pastalan include *University-Linked Retirement Communities: Student Visions of Eldercare* and *Housing Choices and Well-Being of Older Adults: Proper Fit*. He co-edited with Ruth Brent the books *Popular American Housing: A Reference Guide* and *Aging Autonomy and Architecture: Advances in Assisted-Living*. His other publications include *Assisted Living: Sobering Realties* and numerous articles in various academic and professional venues.

About the Contributors

Susana Martins Alves, PhD, is a postdoctorate fellow in the department of Psychology at Universidade Federal do Rio Grande do Norte in Brazil. She completed her undergraduate degree in Psychology and Master's in Environmental Psychology. She earned a PhD in Architecture at the University of Wisconsin-Milwaukee in Environment-Behavior Studies. Her research interests focus on perception of built and natural environments, restorative experiences and design of therapeutic settings, aging environments, and research methods in Environmental Psychology. She will start working as a postdoctoral research fellow in OPENspace Research Centre at Edinburgh College of Art.

Anna Bengtsson, MSc (in Landscape Architecture), is a Doctoral Candidate and Lecturer in Health & Recreation at the Department of Landscape Planning, Swedish University of Agricultural Sciences, Alnarp. Her main area of research interest is on the meaning of outdoor environment design to older persons in need of care. As a lecturer she is working mainly with courses focusing on the importance of the outdoor environment to health and well-being.

Robert D. Brown, PhD, is Professor of Landscape Architecture in the School of Environmental Design and Rural Development, University of Guelph, Canada. He holds a doctorate in Micrometeorology and a master's in Landscape Architecture, from the University of Guelph. Professor Brown is on the Editorial Board of the academic journals *Landscape and Urban Planning*, *Landscape Research*, and *Sustainability Science* and is an active contributor to professional and scholarly publications. He is also the Chair of the Canadian Academic Council of Landscape Architecture. He is co-author of *Microclimatic Landscape Design: Creating Thermal Comfort and Energy Efficiency* published by John Wiley & Sons, and co-editor of *Satoyama: The Traditional Rural Landscape of Japan*, published by Springer.

Gunilla Carlsson, PhD, and Reg. Occupational Therapist, is a research associate at the Department of Health Sciences at Lund University, Sweden. For the past several years, she has had close collaboration with the Department of Technology and Society, Lund University, and she is involved in research focusing on persons with long-term somatic illness and functional disabilities at the Vårdal Institute, Lund University, Sweden. Her main research interest is participation in activities outside the home, with a focus on accessible and usable environments for people with functional limitations.

Uriel Cohen, DArch, has been engaged in architectural education, research and practice since 1972. His main interests include: programming for architectural design; user-based research information and its translation into design guidance; environments for specialized user groups; health care and acute care environments; and research-based design innovations. Dr. Cohen holds a certificate in Gerontology and Design from the University of Michigan, and is a Fellow of the Gerontological Society of America; his most recent research and design applications focus on environments for older persons and in particular those with Alzheimer's disease and related dementias. Two of his books about aging and environment, *Holding on to Home: Designing Environments for People with Dementia*, and *Contemporary Environments for People with Dementia* are both published by Johns Hopkins University Press. Cohen is the co-recipient of five *Progressive Architecture* citations and awards for his work in the field of applied architectural research. Cohen's professional practice, primarily in design for aging, includes over two hundred programming and design consultations projects across the USA and in Canada.

Galen Cranz, PhD, is Professor of Architecture, a sociologist by training, who teaches social and cultural approaches to architecture and urban design at the University of California at Berkeley. She has studied housing for the elderly for the State of New Jersey and the City of San Francisco and published several articles stemming from this early research. Her current research, about how aesthetic taste affects interior design and other places, is the culmination of a 1975 study of aesthetic choices made by residents in publicly assisted housing for the elderly in New Jersey. She is also the author of the now classic *The Politics of Park Design: A History of Urban Parks in America* and continues to have a strong interest in the use of open space. She is also a certified teacher of the Alexander Technique, a somatic perspective fundamental to her quietly radical book *The Chair: Rethinking Culture, Body and Design.* Her research for *The Chair* has earned the 2004 Achievement Award from the Environmental Design Research Association, and extensive media coverage nationally and internationally. She is a Principal Investigator for the 2005-2007 AIA Latrobe Fellowship to explore and create evidence-based design for Kaiser Permanente Hospitals.

Lois J. Cutler, PhD, is a research associate in the School of Public Health at the University of Minnesota. She holds a doctorate in housing from the University of Minnesota and completed a National Institute of Aging post doctoral fellowship in gerontology. Dr. Cutler has focused her research on physical environments of long-term care settings; specifically looking at the interaction between the users of the environment, the characteristics present in their environment, and the outcome of that interaction. She developed assessment protocol for evaluating functional and home-like characteristics of assisted living environments and was part of a quality of life study that developed theoretically derived procedures to assess and compare environments in nursing homes at three nested levels: resident rooms, the nursing unit, and the overall facility. She is involved with the development and design of HUD 202 independent senior housing.

Maria Vittoria Giuliani is a senior researcher at the Institute of Cognitive Science and Technology (ISTC) of the Italian National Research Council (CNR) in Roma. She actively participates in the International Association for People-Environment Studies. She is a permanent member of the Board of Referees of the Italian Ministry for the University and Research. She has been involved in environmental psychology for the last twenty years and is responsible for many research projects. She is affiliated lecturer and member of the doctoral programme in social psychology at the University of Rome "La Sapienza." One continuing line of research concerns users' residential needs and satisfaction across the lifespan. Recently, this perspective has been widened, focusing on assistive technology for elderly people both at home and in sheltered housing, and on strategies for improving communication with patients with sensorial disabilities in hospitals. Another important line of research concerns the concept of attachment to place and psychological consequences of mobility.

Patrik Grahn, MSc Agr Dr, is Professor in landscape architecture and environmental psychology at the Swedish University of Agricultural Sciences, Department of Landscape Planning, Alnarp, where he is Head of the subdivision Health & Recreation. He holds an MSc Agr Dr, and Biologist MSc. He is a member of the Scientific Swedish Royal Academy of Forestry and Agriculture. His research is focused on how to plan and design environments where people can stay healthy and/or grow healthy, and can develop both mentally and physically. Some projects focus on ordinary people in their everyday life at work and at home, and some projects have a special focus on children, old people and the ill. Most of his research projects are multidisciplinary: landscape architecture, horticultural therapy, environmental psychology, medicine, physiotherapy and psychology.

The Role of the Outdoors in Residential Environments for Aging

Gowri Betrabet Gulwadi, PhD, is Assistant Professor of Interior Design in the Department of Design, Textiles, Gerontology, & Family Studies at the University of Northern Iowa in Cedar Falls. Her research interests in supportive environments for older adults, restorative aspects of natural and built environments, and coping strategies of people under stress were further shaped by experiences during her PhD in Architecture: Environment-Behavior Studies from the University of Wisconsin-Milwaukee. Her current research includes caregiving experiences within home environments and stress-reducing aspects of school environments. She is a Co-Principal Investigator with Dr. Maggie Calkins in a study funded by the Coalition for Health Environments Research (CHER) exploring the environmental correlates of falls in health care environments. Her teaching emphasizes the integration of environment-behavior issues in the programming and design of sustainable spaces.

Lauren Harris-Kojetin, PhD, is Director of Research at the Institute for the Future of Aging Services, an applied research center at the American Association of Homes and Services for the Aging (AAHSA) in Washington, DC. She oversaw the survey of AAHSA-member continuing care retirement communities (CCRCs) that is the basis for the article on which she is a co-author in this book. Dr. Harris-Kojetin oversees the applied research and evaluation agenda of the Institute, collaborates with aging services providers to evaluate promising practices they implement, and encourages aging services providers to implement evidence-based models of care and service. She has over 17 years of experience in applied research with an emphasis on: health care quality for older adults; older consumer health care decision making; and program evaluation in aging services settings. Dr. Harris-Kojetin publishes regularly on the implications of applied research findings for providing improved quality of care and services for older adults. She has a PhD and MA in Public Policy from Rutgers University.

Anjali Joseph, PhD, is the Director of Research at The Center for Health Design based in Concord, California. She recently completed her PhD in Architecture at the Georgia Institute of Technology, Atlanta, Georgia. She is trained as an architect and has her Masters in Architecture from Kansas State University. Her research focuses on the relationship between the design of the physical environment and health.

Rosalie A. Kane, PhD, is Professor in the School of Public Health, the School of Social Work, and the Center for Biomedical Ethics, all at the University of Minnesota. She holds a doctorate in social work from University of Utah and a master's degree from Simmons College School of Social Work in Boston. For many decades, she has conducted large-scale research related to long-term care for seniors and other people with disabilities, including nursing home care, assisted living, and home care; she was PI for a 5-year CMS-funded study to develop measures and indicators of quality of life for nursing homes. She is a prolific author of books, monographs, and peer-reviewed articles. Dr. Kane co-authored the widely-used book *Assessing Older Persons: Measures, Meaning, and Practical Applications*.

Anne Kearney, PhD, is Affiliate Assistant Professor at the University of Washington, College of Forest Resources and a consultant in environmental psychology. She holds a BA in Psychology/Cognitive Science from Stanford University and an MS in Resource Policy and Behavior and Ph.D in Environmental Psychology from The School of Natural Resources and Environment, University of Michigan. Her current research focuses on the impacts of urban environments on well-being.

Kristen M. Kiefer, MPP, is the Research Manager for the Access to Benefits Coalition (ABC) at The National Council on the Aging. At the Coalition, Ms. Kiefer develops and implements data collection strategies to identify and evaluate cost-effective and efficient practices to enroll Medicare beneficiaries into public benefits. She also provides training and technical assistance on the Extra Help to be provided people with limited means through the new Medicare Prescription Drug Coverage. Prior to joining the Coalition, Ms. Kiefer served as a Research Associate at the Institute for the Future of Aging Services in Washington, DC. She received a Master of Public Policy degree from Georgetown University and a Bachelor of Science in Health and a Bachelor of Business Administration from Ohio University.

Clare Cooper Marcus is Professor Emerita in the Departments of Architecture and Landscape Architecture at the University of California, Berkeley. She is the author/coauthor/editor of several books including *Housing as if People Mattered, People Places, House as a Mirror of Self*, and *Healing Gardens*. She speaks internationally on the topic of outdoor space in healthcare facilities, and is the Principal of Healing Landscapes, Berkeley, CA.

Johan Ottosson, Agr lic, landscape architecture, MScHort, is a Doctoral Candidate in Health & Recreation at the Department of Landscape Planning, Swedish University of Agricultural Sciences, Alnarp. His main area of research interest is the role of nature in people's rehabilitation and coping with crises. Of special interest in his research has been that of elderly people and people suffering from crises. His most noted work and part of his doctoral thesis is an introspective study of the author's own experiences and perceptions of nature during a period of rehabilitation following a traumatic head injury. (The Importance of Nature in Coping with a Crisis (Ottosson, 2001)/Naturens betydelse i en livskris (Ottosson, 1997). He is a frequent and popular lecturer.

Susan Rodiek, PhD, holds a doctorate in Architecture from Cardiff University. She is Assistant Professor of Architecture and Associate Director of the Center for Health Systems & Design at Texas A&M University. Dr. Rodiek is a nationally-registered architect with a broad range of experience in building and community projects. Her research is focused on behavioral implications of the design of healthcare and long term care settings, with an emphasis on the therapeutic potential of access to nature and the outdoors.

Massimiliano Scopelliti, PhD, holds a doctorate in Social and Environmental Psychology, and currently works at the Institute of Cognitive Science and Technology (ISTC) of the Italian National Research Council (CNR) in Rome. As a guest professor he teaches Social Psychology at the University of Rome "La Sapienza." He is a member of the International Association for People-Environment Studies (IAPS). His main research interest is in environmental psychology. Specific research topics are: the evaluation of the restorative potential of natural and built environments; the use/acceptability of new technologies by elderly people performing everyday activities in the domestic environment; the evaluation of urban quality and residential satisfaction; the assessment of environmental humanization and legibility in the hospital environment; the perception of environmental quality and satisfaction in the workplace.

Takemi Sugiyama, PhD, received his Master of Architecture from Virginia Tech, and PhD in Environment-Behaviour Studies from the University of Sydney, Australia. He has been working with the OPENspace Research Centre at the Edinburgh College of Art, Scotland, on a project that examines the relationship between neighbourhood environments and older people's outdoor activity and their Quality of Life in the UK. In the future he will be working with the School of Population Health, The University of Queensland, Australia, to study environmental determinants of physical activity.

Joyce Tang, MLA, is currently living and practicing landscape architecture and planning in Calgary, Alberta, Canada. She was born and raised in Toronto, Ontario, Canada. She obtained her Bachelor of Urban and Regional Planning from the University of Waterloo, and her Master of Landscape Architecture degree at the University of Guelph. Her areas of interest include research and design of spaces that improve the health of the user. Currently, she is working to expand her research in order to build awareness of the positive impact that well landscaped spaces and healing gardens can bring to health care facilities. She aims to do so through the scientific collection of data characterizing the relationship between the physiological health of people and healing gardens.

Catharine Ward Thompson is Research Professor of Landscape Architecture at Edinburgh College of Art. She was educated at Southampton and Edinburgh Universities. Since 2001 she has been Director of OPENspace– the research centre for inclusive access to outdoor environments–based at Edinburgh College of Art and Heriot-Watt University. She is a Fellow of the UK Chartered Landscape Institute and the Royal Society for the encouragement of Arts, Manufactures and Commerce (RSA), an adviser on research for the Forestry Commission and the Engineering and Physical Sciences Research Council, and a member of the Scottish Physical Activity and Health Council, advising the Deputy Minister for Health and Community Care. She is principal investigator for the Engineering and Physical Sciences Research Council consortium project, I'DGO–inclusive design for getting outdoors–which aims to identify effective ways of ensuring that the outdoor environment is designed inclusively, to improve the quality of life of older people.

Daniel Winterbottom, MLA, is Associate Professor at the University of Washington, College of Architecture and Urban Planning, Department of Landscape Architecture. He holds a Bachelor of Fine Arts from Tufts University and a Master of Landscape Architecture from the Graduate School of Design, Harvard University. Mr. Winterbottom continues his research in the area of therapeutic gardens and writes and lectures widely on the topic. In his professional and academic work he has designed many therapeutic gardens in prisons, AIDS facilities, cancer support facilities and for disabled children and adults. He is the principal of Winterbottom Design Inc., a firm specializing in the design of therapeutic gardens.

Charlene Young, BArch, is a 2004 graduate of the Department of Architecture, University of California, Berkeley, and is currently working in affordable housing with Devine & Gong, Inc. in San Francisco.

Craig Zimring, PhD, is an environmental psychologist and professor of architecture and of psychology at Georgia Tech. His work focuses on understanding the relationships between the physical environment and human satisfaction, performance and behavior. He has conducted studies for the Robert Wood Johnson Foundation, US General Services Administration, Santa Clara County's Valley Medical Center, Ministry of Education of France, World Bank, California Department of Corrections, US Courts, US Department of State, Florida Board of Regents, California Department of General Services, Army Corps of Engineers and many others. His projects have included numerous post-occupancy evaluations of healthcare facilities and other buildings, development of programming and architect selection processes, development of on-line evaluation databases, design guides and other tools, guides and policy analyses. He has served on the board of several organizations including the Robert Wood Johnson Foundation's *Building Bridges* program, National Research Council's Board on Infrastructure and the Constructed Environment, the Environmental Design Research Association and others. He has won 10 awards for his research.

Foreword

It is barely four decades since the first studies began to appear in the field known variously as environment-behavior studies, social factors in design, and environmental psychology. The initial research in these areas used relatively simple methods in trying to establish a basic understanding of how the environment influences human perceptions sophisticated and studies started to consider how interactions with the physical environments might vary, for example, by age, gender, socio-economic and ethnic/racial background.

The criticism that there was a significant "gap" between the methods and thinking employed by social science researchers and those employed by design processionals began to be addressed as some studies concluded with sets of design implications, and books of design guidelines based on research findings started to appear.

If this latter development might be termed state two of this field, the publication of this book marks a transition to stage three: a collection of research papers addressing a specific age group in a particular environment. This is the first scholarly publication that considers in depth the use of the outdoors by older adults, the benefits in terms of health and well-being, and the various barriers–physical, physiologic, psychological, administrative–to that use.

Authors come from a variety of academic disciplines and national backgrounds, investigate outdoor space use in a variety of settings, and employ a range of methods and research designs. this is indeed a benchmark volume, particularly pertinent at this time when the elderly component of the population is growing in most Western countries and

[Haworth co-indexing entry note]: "Foreword." Marcus, Clare Cooper. Co-published simultaneously in *Journal of Housing for the Elderly* (The Haworth Press, Inc.) Vol. 19, No. 3/4, 2005, pp. xxix-xxx; and: *The Role of the Outdoors in Residential Environments for Aging* (ed: Susan Rodiek, and Benyamin Schwarz) The Haworth Press, Inc., 2005, pp. xxiii-xxiv. Single or multiple copies of this article are available for a fee from The Haworth Document Delivery Service [1-800-HAWORTH, 9:00 a.m. - 5:00 p.m. (EST). E-mail address: docdelivery@haworthpress.com].

Available online at http://www.haworthpress.com/web/JHE
© 2005 by The Haworth Press, Inc. All rights reserved. *xxiii*

the health of this cohort is critically important to families, healthcare organizations, and the providers of residential settings.

This book can be seen as a snapshot of where we are right now in this important field of inquiry. These studies will inspire other researchers to delve further into this important topic, offer needed guidance to the designers of outdoor space, and provide firm support to the administrators and staff of settings that cater to the needs of the elderly.

Clare Cooper Marcus

Introduction:
The Outdoors as a Multifaceted Resource for Older Adults

The idea of the outdoors as an important resource for older adults is not new, but recently it is receiving increased research attention. Growing awareness of the physical environment as a therapeutic intervention is contributing to a critical re-examination of how the different facets of the residential environment have the capacity to promote or detract from wellness (e.g., Cutler, 2000; Brownson et al., 2001; Frank et al., 2005). Research increasingly suggests that older adults may benefit from contact with nature and the outdoors, in ways that are similar to other age groups, as evidenced by increased satisfaction and reduced psychological and physiological stress (e.g., Browne, 1992; Rodiek, 2002; Talbot & Kaplan, 1991). Several pathways have been identified by which benefits may occur, such as improved sleep and increased physical activity, some of which are explored in this volume.

Coming from a different perspective, consumer-driven market demands are also beginning to affirm the outdoors as an essential element in planning housing for older adults. For example, a recent study published by the National Association of Home Builders (Wylde, 2002) found that "the top six amenities that would influence the 55+ homebuyer to move to a specific community are *walking and jogging trails, outdoor spaces,* public transportation, *open spaces, a lake and an outdoor pool*" (Mahmood, 2002, p. 5, italics added). There is also evidence that providers are increasingly recognizing the value of including

[Haworth co-indexing entry note]: "Introduction: The Outdoors as a Multifaceted Resource for Older Adults." Rodiek, Susan, and Benyamin Schwarz. Co-published simultaneously in *Journal of Housing for the Elderly* (The Haworth Press, Inc.) Vol. 19, No. 3/4, 2005, pp. 1-6; and: *The Role of the Outdoors in Residential Environments for Aging* (ed: Susan Rodiek, and Benyamin Schwarz) The Haworth Press, Inc., 2005, pp. 1-6. Single or multiple copies of this article are available for a fee from The Haworth Document Delivery Service [1-800-HAWORTH, 9:00 a.m. - 5:00 p.m. (EST). E-mail address: docdelivery@haworthpress.com].

Available online at http://www.haworthpress.com/web/JHE
© 2005 by The Haworth Press, Inc. All rights reserved.
doi:10.1300/J081v19n03_01

outdoor elements as an essential element in housing for the elderly. For example, a recent survey of professionals in dementia-serving assisted living found that industry professionals placed "secure outside activity space" near the top of the list of the most important elements in the physical design of their communities (Keane et al., 2003, p. 15). The growing awareness of the value of outdoor elements begins to confirm what many have long suspected: the outdoors is a highly desired and potentially valuable resource for older people.

In the lives of those who move from independent community-dwelling to living in a residential care facility, the role of the outdoors may undergo a dramatic shift. For many, their connection with the outdoors may greatly diminish or cease altogether (Stoneham & Jones, 1997), as the outdoor features and activities they previously enjoyed, such as a favorite patio, porch, or garden, may no longer be available. In addition, progressive physical decrements make it increasingly harder to navigate the outdoor environment, which is typically more challenging than being indoors, in terms of walking surfaces, temperature, accessibility, etc. For these two reasons–the loss of previously well-used spaces, and the need for supportively-designed elements–it becomes increasingly crucial for designers to carefully identify and address the actual needs of older adults for outdoor space, in a variety of residential settings.

Preliminary indications are that going outdoors may contribute to wellness, and that a majority of long-term care residents have an interest in spending time outdoors, and perceive benefits to their health and well being (e.g., Browne, 1992; Rodiek, 2003). In spite of this, the widespread perception is that many facility residents do not spend as much time outdoors as might be desirable (e.g., Hiatt, 1980; Heath & Gifford, 2001). The outdoor environment has a peculiar role in the overall residential environment, in that, unlike most indoor areas such as living rooms, corridors, and individual units, spending time outdoors is nearly always entirely *optional*. Many long-term care residents seldom go outdoors at all (see Cutler & Kane, Kearney & Winterbottom, this volume), and the question arises: are they missing an important ingredient that could contribute to their health and well-being? Although for certain residents, contact with the outdoors would be harmful rather than beneficial (such as those with severe asthma), the majority of older adults appear to be able to go outdoors, and are fairly aware of health benefits (Rodiek, 2003; Kono et al., 2004).

The papers in this special volume address this important topic on a number of different levels. Kearney and Winterbottom discuss their study of forty residents in three long-term care facilities who were interviewed about the significance of outdoor green spaces in their close environment. The findings indicate that in spite of the fact that residents place a high value on access to green spaces they spent relatively limited time in the outdoors. The limitations for greater use of green spaces include physical barriers, lack of staff assistance, and design configuration problems. Cutler and Kane reiterate these findings in a much broader study in which the investigators collected data from 1,988 residents of 131 nursing home units in 40 nursing homes located in 5 states. The researchers found that the majority of the reviewed units had limited outdoor facilities and 32% of the physically able residents used the outdoors less than once a month. In the third article in this collection, two researchers from Sweden, Bengtsson and Carlsson, report on their study regarding the use of outdoor spaces by residents of three nursing homes. Their qualitative research that was based on focus groups with staff members elicited several themes that the authors grouped into two primary and ten secondary themes with implications for inspiring design of the outdoors.

Cranz and Young who studied gardens of a housing project affiliated with the University of California at Berkeley discuss the problem of underutilized outdoor spaces again. The researchers recently conducted a follow-up set of interviews and observations to a post occupancy evaluation study originally dated to 1994. Their article explores reasons for the underused spaces and makes recommendations for design interventions for the site. The suggestions may have implications for other outdoor spaces in housing for the elderly. Rodiek discusses the features of outdoor environments that attract or deter usage. She reports on her study in fourteen assisted living arrangements with 108 residents and concludes that accessibility, overall aesthetics and specific features, such as shade, sitting accommodations, plants and views influenced the use of outdoor environments.

Alves, Gulwadi and Cohen used six photographs of natural settings to elicit nature-related responses of preference among Hispanic and Anglo-American elderly people. Their findings show clear association of outdoors activities among older adults of particular culture. The authors follow the research report with culture specific recommendations for programming and design of outdoor settings.

The piece written by Joseph, Zimring, Harris-Kojetin, and Kiefer looks into another aspect of the role of the outdoors for older adults. The authors examined how the presence and visibility of outdoor and indoor physical activity resources influence participation in physical activity. The investigators surveyed 800 continuing retirement communities and other forms of housing projects for older adults in their effort to identify the reasons older people engaged in outdoor and indoor physical activities. Their findings indicate that campuses with more attractive facilities have higher participation of older adults in various types of physical activities. Sugiyama and Ward Thompson conducted small scale studies in an attempt to develop a model that uses "environmental support" as a predictor of quality of life for elderly people. They defined environmental support as the degree to which the outdoor environment facilitates individual's projects that involve outdoor activities. The investigators found the correlation between quality of life and the evaluation of nearby natural environments such as gardens, local parks, and pedestrian paths was relatively high. The authors discuss the implications of their model and future research directions.

"The Effects of Viewing a Landscape on Physiological Health of Elderly Women" by Tang and Brown measure blood pressure and heart rates of elderly women who spent time looking at a natural landscape. They discovered that viewing the natural landscape resulted in lower systolic and diastolic blood pressures and lower heart rates of women who were watching the natural landscape, in comparison to the women in the control group who had no view of the outdoors. The authors speculate that the results have important implications for the design of housing for the elderly. Scopelliti and Giuliani studied 192 elderly people in Rome, Italy for their restorative experiences in natural and built environments. The authors note "restorativeness emerges as a consequence of complex person-environment transactions, in which place-specific processes may occur." The article deals with theoretical and practical implications of the findings from an environmental psychology perspective.

In the final paper, Ottosson and Grahn report on an intervention study they conducted in Sweden, in which 15 elderly people were tested for the power of concentration, blood pressure and heart rate before and after an hour of rest in a garden and in an indoor setting respectively. "The results indicate that power of concentration increases for very old people after a visit in a garden" in comparison to resting in their favorite room. In spite of the small sample, the authors believe in the implications of the study for the vulnerable population of older adults who have limited access to outdoor environments.

The famous landscape architect Luis Barragan once said that "... beauty speaks like an oracle, and man has always heeded its message in an infinite number of ways ... a garden must combine the poetic and the mysterious with serenity and joy." It is the multifaceted resource of the outdoors with the mystic and the beauty of nature that we seek to attain in our efforts to provide better places for older adults.

Susan Rodiek, PhD, NCARB
Benyamin Schwarz, PhD
Editors

REFERENCES

Browne, C. A. (1992) The role of nature for the promotion of well-being of the elderly. In *The Role of Horticulture in Human Well-Being and Social Development: A National Symposium* (ed. Relf, D.) Timber Press, Portland, OR, pp. 75-79.

Brownson, R. C., Baker, E. A., Housemann, R. A., Brennan, L. K. and Bacak, S. J. (2001) Environmental and policy determinants of physical activity in the United States, *American Journal of Public Health*, (91), 1995-2003.

Cutler, L. J. (2000) Assessment of physical environments of older adults. In *Assessing Older Persons: Measures, Meaning, and Practical Applications* (eds. Kane, R. L. and Kane, R. A.) Oxford University Press, New York, pp. 360-379.

Frank, L. D., Schmid T. L., Sallis, J. F., Chapman, J., Saelens, B. E. (2005) Linking objectively measured physical activity with objectively measured urban form: Findings from SMARTRAQ, *American Journal of Preventive Medicine*, (28), 117-125.

Heath, Y. and Gifford, R. (2001) Post-occupancy evaluation of therapeutic gardens in a multi-level care facility for the aged, *Activities, Adaptation & Aging*, (25), 21-43.

Hiatt, L. G. (1980) Moving outside and making it a meaningful experience. In *Nursing Homes: A Country and Metropolitan Area Data Book*. U.S. Government Printing Office: Public Health Services Publication #2043. National Center for Health Statistics, Rockville, MD.

Keane, W. L., Cislo, A. and Fulton, B. R. (2003) Defining the dementia market, *Assisted Living Today*, (10), 14-17.

Kono, A., Kai, I., Sakato, C. and Rubenstein, L. Z. (2004) Frequency of going outdoors: A predictor of functional and psychosocial change among ambulatory frail elders living at home, *Journal of Gerontology: Medical Sciences*, (59A), 275-280.

Mahmood, A. (2002) A book for home builders interested in the 55+ market, *Seniors' Housing Update*, (11), 5-6.

Rodiek, S. D. (2002) Influence of an outdoor garden on mood and stress in older persons, *Journal of Therapeutic Horticulture*, (13), 13-21.

Rodiek, S. D. (2003) How the built environment affects outdoor usage at assisted living facilities for the frail elderly. In *Proceedings of the Research Student Conference: Welsh School of Architecture*, Cardiff University, Cardiff, Wales, pp. 66-78.

Rodiek, S. D. (2004) Therapeutic Potential of Outdoor Access for Elderly Residents at Assisted Living Facilities, Unpublished doctoral dissertation, Architecture, Welsh School of Architecture.

Stoneham, J. and Jones, R. (1997) Residential landscapes: Their contribution to the quality of older people's lives, In *Horticultural Therapy and the Older Adult Population, Part 1*(ed. Wells, S. E.) Haworth Press, New York, 17-26.

Talbot, J. F. and Kaplan, R. (1991) The benefits of nearby nature for elderly apartment residents, *International Journal of Aging and Human Development*, (33), 119-130.

Wylde, M. (2002) *Boomers on the Horizon: Housing Preferences of the 55+ Market*, BuilderBooks (National Association of Home Builders), Washington, DC.

Nearby Nature and Long-Term Care Facility Residents: Benefits and Design Recommendations

Anne R. Kearney
Daniel Winterbottom

SUMMARY. As the population in North America continues to age, long-term care facilities for housing the elderly are likely to become even more important. Because one of the primary foci of these facilities is on sustaining and enhancing quality of life while eventually helping patients and families cope with the dying process, both the physical and social environments are critical to the facilities' success and the users'

Dr. Anne R. Kearney is Affiliate Assistant Professor, University of Washington, College of Forest Resources and a private consultant in environmental psychology. She holds a BA in Psychology/Cognitive Science from Stanford University and an MS in Resource Policy and Behavior and PhD in Environmental Psychology from The School of Natural Resources and Environment, University of Michigan.

Daniel Winterbottom is Associate Professor, University of Washington, College of Architecture and Urban Planning, Department of Landscape Architecture. He holds a Bachelor of Fine Arts from Tufts University and a Master of Landscape Architecture from the Graduate School of Design, Harvard University.

Address correspondence to: Anne Kearney, College of Forest Resources, Box 352100, University of Washington, Seattle, WA 98195-2100 (E-mail: anne@alexanne. net).

This work was supported by the USDA Forest Service National Urban and Community Forestry Advisory Council (NUCFAC). The authors would also like to thank their research assistants, Jan Satterthwaite and Jenna Tilt, who conducted the interviews, and the wonderful staff at the three facilities who assisted them with this research.

[Haworth co-indexing entry note]: "Nearby Nature and Long-Term Care Facility Residents: Benefits and Design Recommendations." Kearney, Anne R., and Daniel Winterbottom. Co-published simultaneously in *Journal of Housing for the Elderly* (The Haworth Press, Inc.) Vol. 19, No. 3/4, 2005, pp. 7-28; and: *The Role of the Outdoors in Residential Environments for Aging* (ed. Susan Rodiek, and Benyamin Schwarz) The Haworth Press, Inc., 2005, pp. 7-28. Single or multiple copies of this article are available for a fee from The Haworth Document Delivery Service [1-800-HAWORTH, 9:00 a.m. - 5:00 p.m. (EST). E-mail address: docdelivery@haworthpress.com].

Available online at http://www.haworthpress.com/web/JHE
© 2005 by The Haworth Press, Inc. All rights reserved.
doi:10.1300/J081v19n03_02

well-being. Healing, or restorative gardens and other designed green spaces have been suggested by many academics and practitioners as important components of these environments, yet there has been relatively little systematic research on the use and benefits of nature in this context. Do elderly residents of long-term care facilities benefit from access to outdoor areas? What are the design characteristics that are most important for this unique population?

Forty elderly residents of three different urban long-term care facilities were interviewed about the importance of outdoor green spaces and views within the facility, their use of the facility's outdoor spaces, benefits they derive from those spaces, and barriers to using the spaces. Facilities differed both in terms of the amount of nature in their outdoor spaces and in the design of, scale of, and access to those spaces. Results show that overall residents place a high value on access to green spaces and derive a number of benefits from these spaces, yet they spend relatively little time in these settings. Barriers to greater use of outdoor spaces included physical limitations, lack of staff assistance, and design issues. Implications for the value of nature spaces in long-term care facilities are discussed, along with specific design recommendations. *[Article copies available for a fee from The Haworth Document Delivery Service: 1-800-HAWORTH. E-mail address: <docdelivery@haworthpress.com> Website: <http:// www.HaworthPress.com> © 2005 by The Haworth Press, Inc. All rights reserved.]*

KEYWORDS. Healing gardens, restorative gardens, nearby nature, long-term care, end-of-life

INTRODUCTION

As the population in North America continues to age, long-term care facilities for the elderly are likely to become even more important. Because the focus of these facilities is on sustaining and enhancing quality of life while eventually helping patients and families cope with the dying process, both the physical and social environments are critical to the facilities' success (Marcus and Barnes, 1999; Spriggs, Kaufman, and Warner, 1998). Healing gardens and other green spaces (ranging from small courtyards to larger forested areas) have been suggested by many as important components of supportive and healing environments (Healy, 1986; Marcus and Barnes, 1999; Spriggs, Kaufman, and Warner, 1998; Tyson, 1998). Despite the widespread recognition among landscape architects, environmental psychologists, and others of the importance of nearby nature both to people's perceptions of the quality of long-term care and, more importantly, to patient quality of life, little empirical research has been done on the benefits of green spaces in these contexts.

In a larger context, there is growing evidence of a beneficial effect of nature on human health and illness recovery (e.g., Moore, 1981; Ulrich, 1984; West, 1986). The long-term care setting is somewhat different, however, as the focus is not on recovery, but on providing an environment that maximizes quality of life and supports residents', care providers,' and loved one's ability to cope with the dying process. Studies have shown that nature may have important benefits in this regard as well. Cimprich (1993) showed a link between restorative nature experiences and enhanced effectiveness in many aspects of life among cancer patients. Such effects have been attributed to recovery and maintenance of directed attention capacity which, in turn, increases the mental resources necessary to cope with an illness (Kaplan and Kaplan, 1989; Kaplan, 1995).

Psychological and Health Benefits of Nature: A Brief Review

Considerable evidence exists supporting the notion that natural environments have significant positive psychological and health benefits (Altman and Wohlwill, 1983; Francis and Hester, 1990; Kaplan and Kaplan, 1989; Relf, 1992; Ulrich, 1999), including a recovery in the capacity to focus attention (Kaplan and Talbot, 1983; Kaplan and Kaplan, 1989), a reduction in stress levels (Ulrich, 1983), and a reduction in blood pressure among older adults (Orsega-Smith et al., 2004). These psychological effects can, in turn, have broad-ranging impacts for both individuals and society. Something as simple as passive interaction with nature through a window view, for instance, has been associated with higher job satisfaction and fewer ailments in the workplace (Kaplan, 1993) and reduced stress levels and ailments in prisons (Moore, 1981; West, 1986).

Sullivan, Kuo, and their students (Kuo and Sullivan, 1996; Taylor et al., 1998) have found wide-ranging impacts of nearby nature on public housing residents. Their study, conducted in the Robert-Taylor housing development in Chicago showed that residents living closer to courtyards with natural elements (e.g., grass, trees), as compared to those living near courtyards that were mostly paved, enjoyed a range of benefits including higher overall satisfaction with their home, better relationships with their neighbors, and a lower level of domestic violence.

Kaplan and Kaplan (1989; Kaplan, 1995) have developed the Attention Restoration Theory to explain these psychological benefits of nature. The theory emphasizes the role of nature in restoring the capacity to focus attention; central to this theory is the concept of *directed attention*. Directed attention allows one to engage in purposeful, directed activity and thought. Directed attention is invoked throughout the day as one is called upon to attend to and deal with the numerous activities that are necessary and important but that are not intrinsically interesting. Directed attention is, however, a limited resource–that is, the underlying cognitive mechanism is subject to fatigue–and is hence subject to depletion. Directed Attention Fatigue (DAF)–something from which everyone has undoubtedly suffered–is characterized by an inability to focus one's attention and inhibit competing thoughts. The impaired cognitive functioning that is the hallmark of DAF may have extensive consequences, including difficulty in planning and decision making, accident proneness, impulsive behavior, increased irritability and depression, and social conflicts. It should be noted that DAF is distinct from stress, although it can both cause and be caused by stress (Kaplan, 1995).

Given the serious consequences of DAF, recovery of directed attention capacity is a very important issue. Sleep is one means of attention restoration, however, when the attentional mechanism is chronically depleted (as might be the case when dealing with situations that make great demands on one's directed attention capacity, such as living in a dangerous inner-city environment or dealing with a terminal illness) sleep may not be enough. Fortunately, restorative activities and environments exist that promote the recovery of directed attention capacity. The Kaplans (1989) have identified four dimensions of restorative environments–fascination, being away, extent, and compatibility–and point out that nature is extremely well-endowed across all these dimensions, making it an ideal restorative environment.

The relationship between natural environments and directed attention capacity has been demonstrated in several studies. Hartig et al. (1991) compared the pre- and post-attentional capacity of three groups: wilderness vacationers, urban vacationers, and non-vacationers. They found that following their vacation, the wilderness group showed significant improvements in attentional capacity (as measured by proofreading performance) while the other two groups showed a post-test decline in attentional capacity. In a more controlled study, Hartig et al. (1991) randomly assigned study participants to one of three groups. All participants completed attention-demanding (i.e., depleting) tasks and then, based on group assignment, spent 40 minutes either walking in a natural environment, walking in an urban environment, or listening to soft music and reading magazines. Again, participants in the nature walk group performed better on the proofreading task, indicating a significant recovery of directed attention.

Passive interaction with nature (as opposed to the more direct interaction studied by Hartig and his colleagues) can also have significant impacts on recovery of directed attention. Tennessen and Cimprich (1995) studied college undergraduates living in dormitories and found that those with more natural views out their windows scored better on a range of attentional measures than those with less natural views.

Benefits of Nature in Health Care Settings

Although the impact of nature on human health and healing is less well-documented than the general psychological benefits of nature, evidence of the powerful healing benefits of nature does exist (for a review, see Ulrich, 1999). In a landmark study, Ulrich (1984) found that post-surgical hospital patients with a view of nature from their hospital window showed faster recoveries and less use of pain medication than those with non-nature views. Rodin and Langer's (1977) work on the health effects of choice and personal responsibility in long-term care settings suggest that in some cases direct interaction with plants may be more beneficial than passive viewing.

Cimprich's (1992, 1993) study of recovering cancer patients supports a link between restorative nature experiences and improved mental functioning, with broad implications. The study showed that post-surgical breast cancer patients suffered from high levels of DAF as compared to non-patient adults. Cimprich hypothesizes that the DAF, in this case, was due to the heavy mental demands of dealing with a life-threatening illness. Study participants were randomly assigned to either an experimental or control group. The experimental group agreed to participate in three restorative activities (of at least 20 minutes each) per week for three months. Many of the activities chosen were nature-related, including walking in nature and gardening. The control group received usual post-surgical care.

Attentional and quality of life measures were done at four time points during the three month study. Results showed significant improvements in attentional performance among the experimental group over the four times they were assessed; the control group showed no improvement. In addition, experimental group participants showed significantly greater increases in quality of life scores by the end of the study than did control group participants. They were also more likely to return to work full time and to start new projects.

While little research exists on the benefits of nature specifically for end-of-life patients, Cimprich's study suggests that such benefits may be far-reaching. Coping with end-of-life issues and coming to terms with the dying process are undoubtedly very stressful and attentionally intensive. In this context, interaction (both active and passive) with the natural environment may play a significant role in improving cognitive functioning, coping and decision-making ability, family and other social relationships, and overall quality of life.

Indeed, there is reason to believe that the benefits of nature extend beyond end-of-life patients to their families and other care-givers. Caring for the special needs of terminally ill patients and coping with patient loss is a very intense activity, requiring great patience and compassion. Presumably, higher levels of attentional capacity would be of considerable help in this context, yet the situation itself likely promotes DAF. In a 1991 study, Canin measured the level of cognitive functioning of AIDS caregivers and looked at the relationship of functioning to the types of leisure activities in which the caregiver participated. She found that nature activities were particularly helpful in promoting robust cognitive functioning and in reducing mental fatigue and the possibility of burnout.

Study Overview

This study will investigate how nearby nature is related to psychological well-being and quality of life of residents in long-term care facilities. Results will provide an understanding of the importance of green spaces to facility residents, the perceived benefits of interaction with these spaces, and the barriers to more frequent use of these spaces. In addition, study results will be used to identify general guidelines for the design of green spaces at long-term care facilities.

METHODS

Structured interviews were conducted with residents of long-term care facilities in the summer and early fall of 2002 in order to explore the perceived benefits of nearby green spaces and barriers to using these spaces.

Site Descriptions

Study participants were selected from three separate urban long-term care facilities in the Seattle metropolitan area: Columbia Lutheran Home, Franciscan Health System Care Center, and Horizon House. Each site is briefly described below.

Columbia Lutheran Home

Columbia Lutheran Home (Columbia) is a 116-bed certified skilled nursing facility offering four levels of care including long-term care, subacute rehabilitation, alzheimer/dementia care, and respite care.

Columbia is located in a residential neighborhood and is surrounded by single family houses and apartment buildings. City views are seen from many of the rooms. Views of the surrounding streets, houses and gardens and an enclosed courtyard make up the other views. Columbia contains two primary gardens. The first, a patio garden, is located just outside the main entrance. Approximately 60′ × 20′ in size, the garden features many planting beds containing perennials, herbs, and annuals, and many urn containers are filled with herbs. Wood benches ring the patio. Access to the space from the residents' rooms is through the front doors, down a covered ramp.

The second space is the interior courtyard. This space, approximately 40′ × 100′, features a serpentine concrete pathway, brick patio, boarder and island planting areas, several seating groupings, container plantings and garden sculptures. The plantings consist of trees, shrub massings, annuals and many containers filled with annuals and seasonal flower arrangements. The space is close to many of the residents' rooms and offers nice views from the abutting rooms.

In addition to these spaces, staff commented that residents are often taken for walks in the neighborhood to see the many vegetated streetscapes and residential gardens.

Franciscan Health System Care Center

Franciscan Health System Care Center (Franciscan) is a 110 bed skilled nursing facility that offers long-term nursing care and short-term treatment for conditions not requiring a hospital stay.

Surrounded by mature conifer trees to the west, a vegetated University of Washington campus to the south, and residences to the east and north, Franciscan offers a range of views of natural scenery depending on the orientation of the room. Views of the surrounding landscapes and the enclosed courtyard are available from many of the resident's rooms. The primary garden is a patio area around which half of the residential units are oriented. The other units, located on the outside of the central hallway, offer views of the woods, the street or of the residences across the street. The courtyard is approximately 40′ × 100′ and contains a concrete serpentine pathway that connects planting beds on one end to a large open paved patio on the other. The planting beds feature a collection of broad leaf evergreen and deciduous shrubs and ground covers. There are no trees. A few containers have annual planting and chairs and tables are located within the patio area and a few chairs flank the sides of the space. Access is from either end of the space and because all the residents' rooms are on the first floor, most should be able to access the space with relative ease.

Horizon House

Located in downtown Seattle, Horizon House (Horizon), a 19-story tower, is a continuing care retirement community offering long-term care as well as 318 independent living apartments and assisted living apartments. Additional services include alzheimer/dementia care and surgical recovery.

Many of the views from the long-term care residents' rooms are of the downtown skyline, the nearby Convention Center, blank rooftops, or a modest-sized park situated below the residential units. Located on a sloping site, the facility contains two main gardens only accessible from within the building. The first is a roof-top garden located above the parking garage three floors below the main entry, and is defined with a series of concrete raised planters with a selection of shrubs and small trees. This space lacks a primary focal point and the concrete walls and planters largely overwhelm the plant materials. The second space is a small gathering space approximately 60′ × 60′ at the lobby level. The ground plane is a lawn with evergreen and deciduous shrub massings, tree plantings, and an arbor with seating underneath. This garden is well loved by many Horizon residents, although access is quite difficult for long-term care residents who must take an elevator, pass through a parking garage, and navigate down to the garden.

Participant Description and Recruitment

Forty residents participated in the study: 13 each from Columbia and Horizon, and 14 from Franciscan. The participant selection criteria were as follows: (1) in long-term care (as opposed to the assisted living or short-term rehabilitation services that a number of the sites offer); (2) appropriate level of cognitive functioning (as determined by facility staff) to be able to answer questionnaire items during one-on-one, in person interview session (no alzheimer patients were included in the study); (3) having lived in facility at least 3 months; (4) English speaking.

A staff member at each facility assisted with identifying residents who met the inclusion criteria. Residents were then approached by staff and invited to participate in the study. All potential participants were approached individually to protect their privacy. Staff members then assisted in coordinating interview times for residents who agreed to participate in the study. All participants were assured of their privacy by both the facility and by the researchers.

Measurement Instruments

Interviews consisted of a series of structured open and close-ended questions designed to assess participants' use of, satisfaction with, and perceived importance of the facility green spaces. Open-ended questions were asked about: (1) favorite and least favorite outdoor areas, including what is liked about those areas and what is done in those areas; (2) perceived benefits of going outside; (3) perceived barriers to going outside; (4) what is liked and disliked about their view from the window; and (5) how much assistance they need in getting around. Likert-scale items (typically on 5-point scales) assessed: (1) how often they've spent time outdoors in the past month (recorded on a 6-point scale from more than once per day to never); (2) importance of having outdoor nature at the facility; (3) satisfaction with outdoor areas; and (4) amount of time spent looking out the window (from 1 = never to 5 = most of the time). In addition, a range of background questions were asked, including age, gender, and length of residence in the facility.

Interviews were conducted in the location of the resident's choice: often in their room and occasionally in a common area. Interviews (including introductions and reading the participant consent form) took approximately 1 hour to complete. Participants were encouraged to take breaks when desired, and several did this. For all Likert-scale questions, participants were shown a large-format laminated scale clearly showing the response choices for that question.

Analysis

Responses to open-ended questions were subjected to content analysis. Percentages of participants mentioning a particular response were calculated for each facility and overall. Qualitative comparisons were then made among the three sites. Comparisions of responses to structured questions were made with the appropriate statistical test (anova or chi-square tests) using a significance level of .05.

RESULTS AND DISCUSSION

Population Description

Across all three sites, residents' average age was 83 years old and ranged from 44 to 101. The average length of stay was 7.67 years. The majority of residents interviewed were female (79.5%). Anova results show differences among the sites both in terms of average resident age ($F(2,29) = 3.89$; $p = 032$) and average length of stay ($F(2,35) = 5.16$; $p = .011$). Tukey post-hoc test results show that Horizon residents were both significantly older than Franciscan residents (means of 90.60 and 77.08 years, respectively) and had lived at the facility significantly longer (means of 7.67 and 1.33 years, respectively). Results are shown in Table 1. Chi-square test results showed no differences in gender distribution among the sites.

Most of the residents interviewed needed at least some assistance in getting around: overall, 59% used a wheelchair and 18% used a walker; no participants were bed-bound. As seen in Table 2, Horizon residents were the most mobile (23% in wheelchairs) while Columbia and Franciscan residents were less mobile (61% and 90% in wheelchairs, respectively).

TABLE 1. Significant Differences Among Sites in Resident Age and Average Length of Residence

	Columbia		Horizon		Franciscan	
	Mean	S.D.	Mean	S.D.	Mean	S.D.
Age	82.33	17.22	90.60[a]	5.74	77.08[a]	9.96
Number of years at facility	3.46	4.42	7.67[a]	7.52	1.33[a]	1.31

[a] Means sharing same superscripts differ significantly at p = .05

Patterns of Outdoor Use

While most study participants spent at least some time outdoors, the frequency of their interactions with outdoor natural settings was relatively low. In response to the question, "About how often have you spent time outdoors in or near nature in the past month?," only 10% of residents overall responded that they went outside at least once a day. Thirty percent (30%) went outside at least once every few days, while 45% went outside less than once a week. Seven and a half percent of residents (7.5%) indicated that they never went out.

Table 3 shows that frequency of outdoor interaction varies across the three sites. Visits are notably less frequent at Horizon where, although the residents are comparatively mobile, access to outdoor areas is the poorest.

Once in the outdoor areas at the facilities, what do residents like to do? Residents were asked to list their favorite outdoor areas at their facility and were then asked what they do in these areas. The responses overwhelmingly indicated that residents tend to engage in passive rather than active nature activities. Forty percent (40%) responded that

TABLE 2. Resident Mobility

	COMBINED	Columbia	Horizon	Franciscan
Wheelchair	59%	61%	23%	90%
Walker	17.9%	8%	37%	8%
No assistance needed	10.3%	15%	15%	0%

TABLE 3. Frequency of Outdoor Area Use

	COMBINED	Columbia	Horizon	Franciscan
More than once per day	2.5%	8%	0%	0%
Once per day	7.5%	8%	0%	14%
Once every few days	20%	39%	15%	7%
Once a week	15%	23%	8%	14%
Less than once a week	45%	15%	77%	43%
Never	7.5%	8%	0%	14%

they observe plants/nature; similarly, 40% responded that they sit and enjoy the space. Twenty percent (20%) of participants liked to watch or interact with people. Tending gardens and reading were each mentioned by 7.5% of residents. Enjoying fresh air, attending social functions, and sleeping were each mentioned by 5% of residents. Results, by site, are shown in Table 4.

Perceived Importance and Benefits of Green Spaces

Is the presence of green space important to residents? If so, what are the perceived benefits of these spaces? Overall, residents indicated that having some outdoor nature at the facility where they live is quite important ($M = 4.58$, $S = .90$ on a 5-point scale). Residents also indicated a high level of satisfaction, overall, with the outdoor areas where they live ($M = 4.47$; $S = .75$). Anova results showed no differences in these ratings among the different sites.

When asked, "What benefits, if any, do you get from going outside?" many residents indicated that the outdoors gives them a chance for a break–they go outside to get fresh air (37.5%) and to feel both mentally (22.5%) and physically (17.5%) better. Other benefits cited included feeling invigorated (15%), getting away from the facility (15%), meeting people (10%), change in attitude or outlook (7.5%), feeling of belonging (7.5%), sun exposure (5%), and general healing effect (5%).

Thirty-eight percent (38%) of Horizon House residents also cited "change in scenery" as a benefit. The fact that so many residents from

TABLE 4. What Residents Do in Their Favorite Outdoor Spaces

	COMBINED	Columbia	Horizon	Franciscan
Observe plants/nature	40%	31%	38%	50%
Sit	40%	31%	38%	50%
Watch/interact with people	20%	38%	8%	14%
Tend gardens	7.5%	8%	8%	7%
Read	7.5%	8%	8%	7%
Fresh air	5%	0%	8%	7%
Social functions	5%	0%	8%	7%
Sleep	5%	15%	0%	0%

Note: Responses included are those listed by at least 5% of residents, overall.

Horizon cited that benefit while none of the residents from the other two facilities did may reflect the limited (and in many cases, lack of) nature views from the Horizon rooms. (Because Horizon House is a 19-story tower situated in the most urban location, views into courtyards or into vegetated residential yards and streets available at the other facilities arc not available here.) Results are shown in Table 5.

Preferred Characteristics of Outdoor Areas

When asked to list their favorite outdoor areas at the facility, responses naturally varied across the three sites. However, when asked what it was about those places that caused them to be preferred, there was strong agreement among residents across the different sites. Plant selection, listed by 32.5% of residents, was the most frequently cited reason for liking the space. Not being indoors was cited by 15% and fresh air was cited by 12.5%. Relaxing/quiet, easy access/negotiable space, and quality of materials were each mentioned by 7.5% of residents while seating places, temperature, and safe were each mentioned by 5% of residents. Responses are shown in Table 6. While there is still much to explore with this population in terms of green space prefer-

TABLE 5. Perceived Benefits from Interaction with Nature Spaces

	COMBINED	Columbia	Horizon	Franciscan
Fresh air	37.5%	38%	54%	21%
Enjoyment/happiness	22.5%	31%	31%	7%
Feel physically better	17.5%	23%	8%	21%
Feel invigorated/stimulated	15%	15%	23%	7%
Getting away from facility	15%	15%	23%	7%
Change in scenery	12.5%	0%	38%	0%
Meet people	10%	15%	15%	0%
Change in attitude or outlook	7.5%	0%	15%	0%
Feeling of belonging/home	7.5%	0%	8%	14%
Tanning/sun exposure	5%	8%	0%	7%
Healing effect	5%	8%	0%	7%

Note: Responses included are those listed by at least 5% of residents, overall.

TABLE 6. Preferred Characteristics of Outdoor Spaces

	COMBINED	Columbia	Horizon	Franciscan
Plant selection	32.5%	23%	38%	36%
Not being indoors	15%	15%	15%	14%
Fresh air	12.5%	8%	15%	14%
Relaxing/quiet	7.5%	15%	0%	7%
Easy access/negotiable space	7.5%	23%	0%	0%
Quality of materials	7.5%	0%	15%	7%
Seating places	5%	8%	8%	0%
Temperature	5%	0%	8%	7%
Safe	5%	15%	0%	0%

Note: Responses included are those listed by at least 5% of residents, overall.

ences, these data begin to suggest some design parameters or guidelines (discussed in the next section).

Barriers to Use of Outdoor Green Spaces

Given that residents place a high importance on having outdoor green spaces at the facilities and that they perceive distinct benefits from the spaces, why aren't residents making more use of outdoor areas? In order to identify barriers to the use of outdoor green space, residents who indicated that they never spend time in the facility's outdoor spaces (7.5% of participants) were asked why. Of those residents who never went outside, all of them (100%) cited their physical condition/lack of mobility. In addition, 33% of residents also indicated that insufficient staff help prevented them from going outside.

All study participants were asked whether they would like to go outside more than they do and, if so, what was keeping them from doing so. Only 20% of residents had no interest in going outside more frequently. Of the other 80% of residents, the inability to access the outdoor spaces–both because of physical limitations (40%) and lack of assistance (20%)–was again the most prominent barrier. In addition, inclement weather/lack of protection from weather was cited as a reason for not going outside more often by 17.5% of residents. Results, by site, are shown in Table 7.

TABLE 7. Perceived Barriers to More Frequent Outdoor Use

	COMBINED	Columbia	Horizon	Franciscan
Physical limitations	40%	46%	46%	29%
Lack of assistance	20%	23%	31%	7%
Inclement weather or lack of protection from	17.5%	31%	8%	14%

Note: Responses included are those listed by at least 5% of residents, overall.

The Importance of the View from the Window

As indicated above, many residents do not go outside as frequently as they would like because of a lack of mobility. For these residents, the interaction with nature may more commonly take place as a visual, rather than a physical, experience. On a typical day, the average resident spends a moderate amount of time looking out a window (in their room or elsewhere) ($M = 3.00$, $S = .1.38$ on a 5-point scale from $1 =$ never to $5 =$ most of the time). Anova results showed no significant differences among residents at the different sites in spite of the differences in views at the facilities. The relative ease of enjoying a view from the window combined with the significant barriers to getting physically outside highlights a significant opportunity for providing residents with a nature experience that may be overlooked at some residences.

Residents were asked what, if anything, they like to look at or watch out the window. Responses were quite similar to the question of what was liked about favorite outdoor spaces; again, nature was a highly valued component. Overall, 35% of residents mentioned views of gardens/plants as desirable. Thirty percent (30%) mentioned views of birds, 22.5% mentioned views of people, and 12.5% mentioned views of landscape/scenery. Views of weather, wildlife/pets, and the sky/atmosphere were each mentioned by 7.5% of participants. Preferred views did not differ significantly among the sites. Results are shown in Table 8.

When asked what, if anything, they disliked about the views out the window, residents generally mentioned built elements and a lack of vegetation. Ten percent (10%) mentioned rooftops as disliked elements, 10% mentioned people, and 8% mentioned houses or buildings. Lack of plants, obstruction/lack of view, and lack of color were each mentioned by 5% of residents.

TABLE 8. Preferred Characteristics of Window Views

	COMBINED	Columbia	Horizon	Franciscan
Gardens/plants	35%	38%	38%	29%
Birds	30%	15%	8%	64%
People	22.5%	23%	15%	29%
Landscape/scenery	12.5%	15%	8%	14%
Weather exposure	7.5%	15%	8%	0%
Wildlife/pets	7.5%	8%	0%	14%
Sky/atmosphere	7.5%	0%	15%	7%

Note: Responses included are those listed by at least 5% of residents, overall.

Not only are nature views preferred by residents, but they potentially offer significant physical and psychological benefits as well. The literature on the benefits of a nature view from the window is growing and shows that nature views are associated with greater mental capacity and better physical health (Kaplan, 1993; Moore, 1981; Tennessen and Cimprich, 1995; Ulrich, 1984; West, 1986).

DESIGN RECOMMENDATIONS

Some attention has been given to design issues in the context of health care facilities (Marcus and Barnes, 1999; Thompson, 2000; Tyson, 1998); however, there has been little systematic study of optimal design characteristics. The data presented here suggest five broad design guidelines, a list which is not meant to be definitive.

Provide Easy Access

The largest barriers to residents' use of outdoor nature spaces were lack of mobility and lack of staff assistance. Some of the gardens at the study sites were difficult to access, requiring extensive navigation through the facility to reach an exit and further navigation once outside to reach the garden. In others, slight grade changes at the garden entry of as little as one step or a 3/4″ threshhold appeared to deter residents from using the gardens.

Access to nature should be provided in a manner that accommodates the range of patient's physical and cognitive abilities. Instead of one access point that may be far from patients' rooms, multiple access opportunities should be created so that patients won't feel the physical effort is too great, lose their stamina, or get disorientated in the process of seeking a physical interaction with nature. The potential for disorientation would be even greater for residents suffering from dementia. A second strategy to increase resident use is to design several smaller garden spaces located throughout the facility instead of a central courtyard. This could reduce the distance from the residents' rooms and help to delineate wayfinding, thereby reducing disorientation.

Efforts should also be made to increase awareness among staff of the benefits of nature interaction and the importance to the patient of such interaction. Such increased awareness may encourage staff assistance in helping residents reach outdoor nature spaces, thereby facilitating resident access. Integrating nature and the use of natural spaces in the facility curricula or activity planning could increase nature interactions among the residents. Care providers in facilities that have integrated nature walks and scheduled social activities in the outdoor spaces in their curricula, indicated that nature interactions occurred quite frequently.

While not specifically addressed in this study, the ability to negotiate within the outdoor space is of equal importance as initial access to that space (Marcus and Barnes, 1999; Thompson, 2000; Tyson, 1998). In addition to eliminating barriers into the outdoor nature spaces, several design recommendations can accommodate users and increase the use of the space. Smooth walking surfaces are critical as many users use wheelchairs, rely on walkers, tend to shuffle or are unsteady when walking. High contrast paving patterns can confuse some users who read the dark pavers as voids and may resist using the pathway. Rails can reassure those that might have limited physical capacity, and frequent opportunities for seating will assist those with limited endurance. Legibility of circulation and visual access to entries and exits is important in larger spaces, especially for users who may suffer dementia, memory loss, or anxiety.

Focus on Vegetation

Residents strongly preferred green spaces that were heavily vegetated, both in terms of visiting outdoor spaces and in terms of the view from the window. This finding was not surprising as a preference for green has been amply demonstrated across a range of populations (see

Kaplan and Kaplan, 1989 for a review). The finding does suggest that gardens should contain a diversity of plants and that, where possible, advantage should be taken of "borrowed" views of trees and other vegetation.

Focus on Passive Nature Interactions

When asked what they liked to do in their favorite outdoor areas, residents overwhelmingly mentioned passive activities such as sitting and viewing nature or wildlife. These preferences likely reflect the physical limitations of many of the residents. Garden design should thus maximize opportunities for passive nature interactions. For example, plants might be selected for visual variety, aroma, or ability to attract wildlife. A water feature would provide both visual and auditory interest. In addition, ample and comfortable seating should be provided and oriented to maximize nature views from within the garden. If possible, some protection from inclement weather should also be integrated into the garden design as lack of protection was frequently cited as a barrier to getting outside more. Shelter from rain, shade from bright sun, and heat lamps in cooler weather may encourage greater use of garden spaces.

Despite the focus on passive activities, opportunities for more active nature interactions should not be ignored. For example, a number of residents indicated that they enjoy tending gardens and the benefits of gardening and horticultural therapy are well documented (e.g., Gabaldo et al., 2003, Simson and Straus,1998, Lewis, 1996; Moore, 1989). Attention must also be paid to other users of the garden spaces–such as facility staff and visitors, including children–who may prefer to engage in more active activities, such as for walking, exercise, or play.

Optimize Window Views

The realities of the physical limitations of residents combined with often over-taxed staff mean that in spite of good design, many residents may not be able to get outside as frequently as they would like. This may be particularly true of residents suffering from dementia, many of whom have limited wayfinding capacity and some of whom (particularly Alzheimer's patients) are not permitted to go outside unattended. In this context, the view from the window is a particularly important opportunity for nature interaction.

As much as possible, windows in the residents' rooms and in common areas should be oriented towards views of nature. These can be borrowed views of existing landscapes or views created though the addition of courtyards, additional tree plantings or, at a minimum, planter boxes.

Take Advantage of Nearby Green Spaces

For many inner city facilities where open space may be limited and often physically and visually removed from the patients, public parks and community gardens can serve as a valuable resource. While access to these places is clearly an issue, providing the transportation and support on a weekly or bi-weekly basis would provide a great benefit for facility residents. Linkages with community service projects at local schools, with park rangers, or with interns may be a viable means to link those residents with public green space.

CONCLUSIONS

As our population ages, it is likely that an increasing number of loved ones will be placed in long-term care facilities. Clearly, quality of life is of one of the most important considerations when choosing a facility. There is a growing body of literature on promoting quality of life at the end of life (Bycock, 1997; Ellis and Abrams, 1994; Lewis, 1982; 1989; 1997; Mandel, 1993; West, 1994; Yamamoto, 1993), and while there has been relatively little research done to date on the benefits of green spaces specifically to this population, there does seem to be an intrinsic understanding of the importance of such spaces. A perusal of facility brochures and web sites shows that the presence of nature is, in fact, an important selling point for many facilities. Many hospice and long-term care facilities have created gardens or integrated the natural grounds into the design of the facility, providing patients with opportunities for both passive and active interaction with nature. In a broader context, many cultures look to nature to provide support, comfort, and a place to gather with friends and family for those dealing with death and dying. It is not uncommon for those facing illness or death to seek natural places that offer settings for contemplative focus and restoration, helping them reestablishing a powerful form of engagement with the "living" world.

The study results presented here provides some empirical evidence that residents of long-term care facilities do have much to benefit from nearby nature and that they place a high value on access to nature. These results, which are based largely on the perceptions and attitudes of long-term care residents, would be complemented by additional research using objective measures of the health and quality of life impacts of nearby nature on both residents and staff. In addition, future research should more explicitly explore the effectiveness of specific design characteristics on preference and well-being. Although such additional research would provide useful information, the case for including green spaces in long-term care facilities is already strong. The potential impacts of nearby nature for this population become even more profound when one considers the relatively low associated cost and very limited side-effects of such a "treatment."

REFERENCES

Altman, I. and Wohlwill, J.F. (Eds) (1983). *Behavior and the natural environment.* New York: Plenum.

Byock, I. (1997). *Dying Well: Peace and Possibilities at the End of Life.* Riverhead Books.

Canin, L. (1991). Psychological restoration among AIDS caregivers: Maintaining self-care. University of Michigan doctoral dissertation.

Cimprich, B. (1992). Attentional fatigue following breast cancer surgery. *Research in Nursing and Health,* 15, 199-207.

Cimprich, B. (1993). Development of an intervention to restore attention in cancer patients. *Cancer Nursing,* 16, 83-92.

Cronbach, L.J. (1951). Coefficient alpha and the internal structure of tests. *Psychometrika,* 16, 297-335.

Ellis, A. and Abrams, M. (1994). *How to Cope with a Fatal Illness.* Barricade Books.

Francis, M. and Hester, R.T. (1990). *The meaning of gardens.* Cambridge, MA: MIT Press.

Gabaldo, M., King, M.D., and Rothert, E.A. (2003). *Health through Horticulture: A Guide for Using the Outdoor Garden for Therapeutic Outcomes.* Chicago: Chicago Botanic Gardens.

Hartig, T., Mang, M., and Evans, G.W. (1991) Restorative effects of natural environment experience. *Environment and Behavior,* 23, 3-26.

Healy, V. (1986). The hospice garden: Addressing the patients' needs through landscape. *The American Journal of Hospice Care,* Nov/Dec, 18-36.

Kaplan, R. (1993). The role of nature in the context of the workplace. *Landscape and Urban Planning,* 26, 193-201.

Kaplan, R. and Kaplan, S. (1989). *The experience of nature: A psychological perspective.* Cambridge: Cambridge University Press.

Kaplan, S. (1995). The restorative benefits of nature: Toward an integrative framework. *Journal of Environmental Psychology,* 15, 169-182.

Kaplan, S. and Talbot, J.F. (1983). Psychological benefits of a wilderness experience. In I. Altman and J.F. Wohwill, Eds., *Behavior and the natural environment.* New York: Plenum, pp. 163-203.

Kuo, F. and Sullivan, W.C. (1996). Do trees strengthen urban communities, reduce domestic violence? *Technology Bulletin 4*, USDA Forest Service, Southern Region, Forestry Report, R8-FR 56.

Lewis, C.A. (1996). *Green Nature/Human Nature*. University of Illinois Press.

Lewis, F.M. (1982). Experienced Personal Control and Quality of Life in Late-Stage Cancer Patients. *Nursing Research*, 31(2).

Lewis, F.M. (1989). Attributions of Control, Experienced Meaning, and Psychosocial Well-being in Patients with Advanced Cancer. *Journal of Oncology*, 7.

Lewis, F.M. (1997). Behavioral Research to Enhance Adjustment and Quality of Life among Adults with Cancer. *Preventive Medicine*, 26.

Lezak, M.D. (1983). *Neuropsychological assessment*. New York: Oxford University Press.

Mandel, S.E. (1993). The role of the music therapist on the hospice/palliative care team. *Journal of Palliative Care*, 9 (4).

Marcus, C. and Barnes, M. (1995). *Gardens in health care facilities*. New York: McGraw Hill.

Marcus, C. and Barnes, M. (1999). *Healing gardens*. New York: John Wiley and Sons.

Moore, B. and Brown, K. (1989). *Growing with Gardening: A Twelve Month Guide for Therapy, Recreation and Education*. University of North Carolina Press.

Moore, E.O. (1981). A prison environment's effect on health care service demands. *Journal of Environmental Systems*, 11, 17-34.

National Hospice Foundation (1999). Web site.

Orsega-Smith, E., Mowen, A., Payne, L., and Goodbey, G. (2004). The interaction of stress and park use on psycho-physiological health in older adults. *Journal of Leisure Research*, 36(2), 232-256.

Relf, D. (Ed.) (1992). *The role of horticulture in human well-being and social development*. Portland, OR: Timber Press.

Rodin, J. and Langer, E. J. (1977). Long-term effects of a control-relevant intervention among the institutionalized aged. *Journal of Personality and Social Psychology*, 35, 275-282.

Simson, S.P. and Strauss, M.C. (1998). *Horticulture as Therapy: Principles and Practices*. Haworth Press.

Spriggs, N., Kaufman, R., and Warner, S. (1998). *Restorative gardens*. New Haven: Yale University Press.

Taylor, A.F., Wiley, A., Kuo, F.E., and Sullivan, W.C. (1998). Growing up in the inner city: Green spaces as places to grow. *Environment and Behavior*, 30(1), 3-27.

Tennessen, C. and Cimprich, G. (1995). Views to nature: Effects on attention. *Journal of Environmental Psychology*, 15, 77-85.

Thompson, J.W. (2000). Healing words: Wither the design of therapeutic gardens? *Landscape Architecture Magazine*, 90(1), 54-57+.

Tyson, M. (1998). *The healing landscape*. New York: McGraw Hill.

Ulrich, R. (1983). Aesthetic and affective response to natural environment. In I. Altman and J.F. Wohlwill Eds., *Human behavior and environment: Advances in theory and research (Vol. 6)*. New York: Plenum, pp. 85-125.

Ulrich, R. (1984). View Through a Window May Influence Recovery from Surgery. *Science*, 224, 420-421.

Ulrich, R. (1999). Effects of gardens on health outcomes: Theory and research. In C.C. Marcus and M. Barnes (Eds), *Healing gardens*. New York: John Wiley and Sons, pp. 27-86.

West, M. J. (1986). *Landscape views and stress responses in the prison environment*. Unpublished master's thesis, University of Washington, Seattle.

West, T.M. (1994). Psychological issues in hospice music therapy. *Music Theory Perspectives*, 12 (2).

Yamamoto, K. (1993). Terminal care and music therapy. *Japanese Journal of Psychosomatic Medicine*, 33 (1).

Yeung, E.W.F., French, P., and Leung, A.O.S. (1999). The impact of hospice inpatient care on the quality of life of patients terminally ill with cancer. *Cancer Nursing*, 22(5), 350-357.

As Great as All Outdoors:
A Study of Outdoor Spaces as a Neglected Resource for Nursing Home Residents

Lois J. Cutler
Rosalie A. Kane

SUMMARY. *Purpose.* Previously, most information on outdoor amenities in nursing homes and the use of outdoor space by nursing home residents has been anecdotal. Using data collected from the Center for Medicare and Medicaid Services (CMS) study on Quality of Life (QOL), this paper describes the availability of outdoor amenities in 40 nursing homes and the resident's perception of their use of that space.

Design and Methods. Resident data were collected from nineteen hundred and eighty-eight residents in 131 nursing units in 40 nursing homes located in 5 states on a broad array of topics including how often they get outdoors and if that amount was as much as they want, too much, or not enough. For each of those nineteen hundred and eighty-eight residents, staff was questioned on how often the resident partici-

Lois J. Cutler, PhD, is Research Associate, Division of Health Services Research, Policy, and Administration, School of Public Health, University of Minnesota, 420 Delaware Street, S.E., D-527 Mayo, Minneapolis, MN 55455 (E-mail: cutle001@umn. edu).

Rosalie A. Kane, PhD, is Professor, Division of Health Services Research, Policy, and Administration, School of Public Health, University of Minnesota, 420 Delaware Street, S.E., D-527 Mayo, Minneapolis, MN 55455 (E-mail: kanex002@umn.edu).

[Haworth co-indexing entry note]: "As Great as All Outdoors: A Study of Outdoor Spaces as a Neglected Resource for Nursing Home Residents." Cutler, Lois J., and Rosalie A. Kane. Co-published simultaneously in *Journal of Housing for the Elderly* (The Haworth Press, Inc.) Vol. 19, No. 3/4, 2005, pp. 29-48; and: *The Role of the Outdoors in Residential Environments for Aging* (ed: Susan Rodiek, and Benyamin Schwarz) The Haworth Press, Inc., 2005, pp. 29-48. Single or multiple copies of this article are available for a fee from The Haworth Document Delivery Service [1-800-HAWORTH, 9:00 a.m. - 5:00 p.m. (EST). E-mail address: docdelivery@haworthpress.com].

Available online at http://www.haworthpress.com/web/JHE
© 2005 by The Haworth Press, Inc. All rights reserved.
doi:10.1300/J081v19n03_03

29

pated in planned outdoor activities. Environmental data were collected using theoretically-derived observational tools that were developed to observe in detail the physical environments experienced by those nineteen hundred and eighty-eight nursing home residents at three nested levels: their rooms (112 items), the nursing unit (140 items); and the facility as a whole (134 items). These analyses focus on the presence or absence of items specific to outdoor space at the unit and facility level.

Results. Descriptive statistics showed great variation in outdoor amenities and access to those amenities across facilities. The majority, 55.7% (n = 73), of the environments of the 131 units had no items featured on the outdoor amenities index. Of the residents who are physically able to go outdoors, thirty two percent do so less than once a month.

Implications. Only recently have the effects of the outdoor environment on well being been systematically studied. This resident-specific data collection on the availability of outdoor amenities and research at the resident level permits hierarchical analysis to examine the effects of outdoor space on resident quality of life. *[Article copies available for a fee from The Haworth Document Delivery Service: 1-800-HAWORTH. E-mail address: <docdelivery@haworthpress.com> Website: <http://www.HaworthPress. com> © 2005 by The Haworth Press, Inc. All rights reserved.]*

KEYWORDS. Outdoor amenities, environmental assessment, resident use of outdoor space

Life outdoors has been an important part of living since the beginning of time. The use of gardens as a place for therapeutic healing can be traced back to ancient Greek, Asian, and Roman cultures where healing temples for their gods were created. The temple for the Greek god Asclepiad (god of healing) was built to include healing gardens, mineral springs and bathing pools as a place for people to come and worship, recreate and heal (Larson & Kreitzer, 2004). Anecdotally, the therapeutic benefits of outdoor space and views of nature (combined with fresh air and exercise) have been widely believed (leading to prescriptions for mountain and beachside cures, pursuit of brisk walks, and small children bundled in snowsuits spending requisite daily time outdoors in winter climates), but only recently have these benefits been systematically studied in healthcare and long-term care settings. Beyond therapeutic benefits, being outdoors arguably is positively associated with improved perceptions of quality of life. Yet outdoor space, outdoor amenities, and access to outdoor space have often been ignored in the design phase or simply value engineered out of a project due to cost when in reality outdoor spaces are especially important to persons sequestered in institutional settings. When outdoor spaces are available to nursing home residents, most often the accessibility and functionality of those spaces are ignored. It is as if they are not considered an integral part of the overall physical environment. Yet we argue that outdoor spaces have the potential of increasing a resident's quality of life and well-being and should be maximized for their potential of providing additional living spaces.

Utilizing findings from a CMS Quality of Life (QOL) study (Kane et al., 2003) that interviewed and assessed the physical environments of 1988 residents in 40 nursing facilities located in 8 states, this article describes the availability of outdoors spaces adjusted by the number of beds in the facility, amenities located in those spaces, the availability of direct access off of a nursing unit, a critique of the outdoor spaces available, and some examples of window views to the outside world that create a sense of being a part of that outside world without leaving the sanctity of the resident's home.

THEORETICAL FRAMEWORK

The Ecological Theory (Lawton & Nahemow, 1973) is as pertinent to outdoor spaces as it is to indoor spaces. Briefly, the theory states that as the aging process continues and the gap between the demands of the environment and the older person's competence widens, there is a loss of mastery over necessary environmental characteristics that can lead to the older person not using or living limited lives in their environments. People's behavior in their environments is directly related to the design of the spaces and that an optimal environment is designed to meet the specific needs and preferences of a person (Cutler, 2000; Kahana, 1975; Christenson, 1990). The ecological model theorizes that behaviors are a function of the interaction of individual factors with the physical, social, psychological, and cultural dimensions of their environment. Behavior and affect are outcomes of a person's level of competence interacting with an environment's level of press. To function at the highest level possible, a person's ability must match demands placed on him or her by their environment. Too little demand results in lack of stimulation, boredom, and even de-conditioning, while too much demand can result in stress and inability to negotiate the environment. The docility hypothesis suggests that the lower level of competence the greater the influence of the environment. Used in isolation, the Ecological Theory emphasizes how the environment stimulates competence and social activities at the expense of considering how the environment fosters other desirable outcomes such as maintaining a sense of continuity, individuality and a sense of place.

Abraham Maslow's (1954) theory of the Hierarchy of Basic Human Needs is a useful conceptual framework for understanding individual needs. Although not originally designed for use in environmental research, it has become a mainstay of housing curriculum as a framework that outlines basic human needs and the way the environment can satisfy those needs both in the workplace and in housing (Denhardt, Denhardt, & Aristigueta, 2002; Lindamood, 1979). The hierarchy is based on the assumption that people are motivated by unsatisfied needs and that certain lower needs need to be satisfied before higher needs can be achieved.

Recently, the Maslow framework was used in a study of the living environment of a continuing care retirement community. Using a predictive model of research, Paul Eshelman and Gary Evans (2002) looked at function and personal meaning as predictors of place attachment and self-esteem. The study determined that once lower level functional needs (physiological and safety needs) are met, both place attachment and self-esteem (social, self-esteem, self-actualization needs) are elevated by environmental features that are preferred by the resident and are a part of favorable memories.

Access to a pleasant and safe outside area in a long-term care facility was found to be a predictor of satisfaction in a study by Greene, Hawes, Wood, and Woodsong (1998). The study looked at how family members define quality in long-term care settings. Six focus groups of family members of people with dementia who were living in or had recently been discharged from an assisted living facility to a nursing home were conducted to gain insights into participants' experiences, perceptions, attitudes about quality and satisfaction. Comments were grouped into four major areas: facility staffing; services; environmental features; and facility operational policies and practices. Environmental features found to be indicative of quality include: a safe environment; access to a pleasant and safe outside area that provides refuge for residents; sufficient space for a range of activities; single-story buildings; a design that did not isolate residents; and personal space that is homelike, clean and allows for personal belongings.

The belief that nature can play a role in quality of life and health care has come into focus since the 1980s when non-traditional therapeutic healing options have become an addition to the traditional medical interventions. In a well documented study of the effects of a window view of nature on outcomes for surgical patients, Roger Ulrich (1984) found that patients with views of nature went home three-quarters of a day sooner, had reduced costs amounting to $500 per case, used fewer heavy medications, had fewer minor complications such as nausea, and exhibited better emotional well-being compared to patients in identical rooms who viewed a brick wall (Berry et al., 2004).

Rachel Kaplan (2001) looked at the psychological restorative benefits of a view from home, finding that windows provide opportunities for prospect-refuge–i.e., a place for viewers to see out (prospect) while at the same time offering a safe place to be (refuge). The penetration of sunlight from windows was related to increased satisfaction and well being, although the study did not determine if the view content affected the outcomes.

Recent attention to the benefits of sunlight has been somewhat over-shadowed by contemporary eagerness to avoid sunlight as a way to limit our chances of skin cancer. But an article in Health Facts (2005) suggests 10 to 15 minutes a day twice a week provides vitamin D synthesis that can provide relief for seasonal affective disorder (depression caused by lack of sunlight), help reset a person's internal clock thus promoting good nighttime sleep and has been found to reduce agitation in residents with dementia. There is also mounting evidence that vitamin D has a role in preventing common cancers, autoimmune diseases, type 1 diabetes, heart disease, and osteoporosis (Holick & Jenkins, 2003). The experience of nursing staff at one nursing facility found that several hours of outdoor activity in the morning greatly reduced unwanted behaviors later in the day and has helped cut its use of psychotropic medications by 40 percent (Gold, 2004).

In their book *Healing Gardens: Therapeutic Benefits and Design Recommendations*, Clare Cooper-Marcus and Marni Barnes (1999) noted that therapeutic landscape design focuses on providing stress relief, alleviation of physical symptoms and overall improvement of quality of life for patients, families and staff. General principles for providing outdoor spaces that focus on improving the overall sense of wellness for the users of that space include (Marcus & Barnes, 1999) (Larson & Kreitzer, 2004):

1. *A Variety of Spaces:* Spaces for individual, group and family use that support solitary use or congregate use. Create spaces that are

easy to access and provide furniture that is supportive of the population using the space.

2. *Prevalence of Natural Green Material:* Reduce the amount of concrete and other hard surfaces.

3. *Encourage Exercise:* Provide a hard surface wandering path at least 3' wide. Example: create a prayer trail with seating along the way.

4. *Provide Positive Distractions:* Examples include a waterfall or small fishpond. Other examples offer opportunity for residents to become involved by providing a raised garden platform or push lawnmowers.

5. *Minimize Intrusions:* Consider placement of outdoor space away from parking lots but yet in a location that is not void of life activity.

6. *Minimize Ambiguity:* Place clearly identifiable features and furniture in the outdoor space for ease of use.

METHODS

Instrument Development

The instruments we used to assess physical environments were developed as part of an overall study of quality of life in nursing homes. One set of study objectives required that we develop and test measures of quality of life as resident outcomes. The resultant measures tap eleven quality of life domains: comfort, security, functional competence, relationships, meaningful activity, enjoyment, individuality, dignity, autonomy, privacy, and spiritual well-being. Each of these outcomes is potentially fostered or impeded by conditions of the physical environment (Kane, 2003; Kane et al., 2003).

To measure physical environments directly, we developed tools that allowed us to assess the unique environments of all residents, beginning with their own dedicated spaces (room and bath), the spaces they share with others on the nursing unit, and the spaces they share with all residents in the facility. For our purposes, we defined environment as "the fixed, semi-fixed, and unfixed components of the physical structure, the furnishings, fixtures, décor, and equipment used." We recognized the diversity of NF residents and the differing preferences and requirements residents might have for their environments, and that some residents would be limited to their near environment because of their frailty. With those goals in mind, we generated a pool of environmental items conceptually associated with resident quality of life (QOL), which pertained to 3 environmental levels: resident room, nursing unit, and facility. Each facility identified the individual nursing units that together comprised the total facility. Generally, a nursing unit means an area within the nursing facility that is served by a single nurses' station. A maximum of 5 nursing units per facility were selected to ensure that there would be at least 10 rooms sampled from each. For larger facilities, a dementia special care unit (SCU) and a Medicare rehabilitation unit were automatically included. Amenities that were not included as part of a specific nursing unit were included in the facility wide assessment. We excluded "backstage" areas such as laundry, staff offices, mechanical rooms and any other spaces that residents did not use in their daily life.

All items were observable and clearly defined, and noted for their presence or absence. The 112-item room and bathroom checklist was applied to each resident in our sample. In addition to outdoor spaces, the nursing unit tool, which included 140 items, looked at nursing stations, corridors, common tub/shower room, lounge and dining spaces, noise, distances and light levels. The goal of assessing outdoor spaces at the unit level was to determine if there was outdoor access directly off the unit and to assess the amenities in that outdoor space. The 134-item facility level tool included all other outdoor and indoor-shared spaces, potentially used by residents, family members, and visitors that were not assessed at the unit level. Any outdoor spaces that were not captured during the unit level assessment were captured at the facility level. Generally, facility level outdoor spaces and amenities were geared towards use by staff, family members and residents who were capable of independently leaving their units.

Field Work

Sample. The environmental assessment checklists were applied in 1999-2000 during the first wave of a national CMS study to develop Quality of Life measures in nursing facilities. The sampling decisions were designed to achieve a sample evenly divided by urban and rural location and size of facility (Kane et al., 2003). Eight facilities were selected in each of 5 states (CA, FL, MN, NJ, and NY) in cooperation with CMS to reflect a range of nursing-home policy environments. In each state in the study, one home was selected as an exemplar from those named in a telephone query of experts who were asked to name a facility in their state with an unusually high QOL. Up to two facilities were randomly selected from a list of nursing homes with 70% or more private rooms. In one state, two facilities with 70% or more private rooms could not be identified. The remaining facilities were slotted by size and rural/urban status. This sampling procedure resulted in a sample that varied greatly in size, privacy, and physical amenities. Thus, the sample for each state should have included several nursing homes likely to offer better-than-standard environments with the majority of homes reflecting the typical standard in the geographic area.

The starting point for the assessment of physical environments was the 1988 individual residents randomly selected for the study. At the individual level, the sample was stratified to be as evenly divided as possible between those functioning at higher and lower cognitive levels using the Minimum Data Set assessment (MDS). The MDS is a comprehensive assessment of each resident's needs, which among other things, assesses the resident's capability to perform daily life functions and significant impairments in functional capacity. The resident sample was further divided among up to 5 units in each facility, and residents in private rooms were over sampled up to 20% of the sample when that was possible. The average number of units per facility was 3 and any dementia SCU or Medicare rehabilitation units were automatically chosen with the other units chosen by random. The procedure yielded 131 distinct nursing unit environments with 21 of those units designated as SCU dementia units.

Data Collection. The room and bath data were collected by 40 research interviewers who also interviewed each resident or their family proxy, interviewed a frontline staff member about each resident, and performed a variety of observational studies. The first author visited all 40 facilities and completed the multiple unit checklists and the overall facility checklists. During that visit, she also conducted a qualitative appraisal of each facility to identify innovative designs as well as noting areas that could be improved upon. After each visit, detailed field notes about each facility were transcribed into a descriptive narrative. The study design also included lengthy interviews with the administrator, the director of nursing, the director of social work, and the director of activities in each facility as part of the search for facility practices that might prove to predict better resident quality of life. We drew on those interviews as well to derive information about policies and practices regarding the use of outdoor spaces.

RESULTS

Description of Sample

In our sample, 18 facilities were 1-story, 7 facilities had 2 stories, another 7 had 3 stories, 7 had 4 stories, and 1 facility was 6 stories high. Of the 131 nursing units, 21 (16%) were designated as special care units. Of those 21 SCU units, 13 (61.9%) have direct access from the unit to outdoor space, 10 of which were located on the ground floor and the other 3 on a second story. Most often the direct access was locked and residents were only able to use the outdoor space when escorted by a staff or family member or on the rare occasion when outdoor activities were scheduled.

Despite the over sampling for facilities with private rooms and residents in private rooms, 1152 (58%) of the 1988 residents were in two-bedded rooms and 256 (13%) were in rooms with 3 or more beds. The room and bath assessments showed that 52% (734) of residents in shared rooms had a view of the outdoors from their bed or their bedside chairs without looking across a roommate's bed. Double-bedded rooms were typically organized so that only one resident had access to the window space. The exception were found in 2 recently renovated facilities where all double rooms were divided with floor-to-ceiling partial walls and each resident had their own window; these semi-private rooms were sufficiently private to permit residents to use completely separate window treatments in their portion of the room. Another issue with windows that we did not set out to measure routinely but that we observed during the visit concerned the height and size of the glass panes. One six-story facility comprised entirely of private rooms organized those rooms around the periphery of the building (20 per floor) and each room had a wide almost floor-to-ceiling window that occupied most of the wall. In contrast, most other facilities utilized higher and smaller windows, cutting off the view from wheelchairs and various locations in the room.

Residents were surveyed as to how often they get outdoors and if that amount is the amount they prefer. Of 1,068 residents who were able to complete the interview and for whom we could construct scales, 334 (32.2%) responded they went outdoors less than once a month, 143 residents (13.4%) responded being outdoors less than once a week, 179 residents (16.8%) about once a week, 169 residents (15.8%) several times a week and 233 residents (21.8%) said they went outdoors every day. When asked if that amount was too much, not enough, or the right amount 692 residents (61%) said it was the right amount; 39% of the residents said it was not the right amount. A distinction was not made between whether the amount of time outdoors was either too much or not enough.

Families were surveyed as to how often their relative gets outdoors and if that amount is the proper amount. Of 1,780 family responses, 769 families (43.3%) responded their relative gets out as much as he/she wants, only 6 families (.3%) responded that amount was too much, 618 families (34.7%) responded the amount was not enough and 387 families (21.7%) responded they did not know.

Staff members were surveyed on how often residents participated in several categories of planned activities. Table 1 shows the frequency of resident participation in three types of planned activities: exercise, social, and outdoor activities. The findings show variation between participation in the activities and especially very limited participation in planned outdoor activities. For 1,988 residents, responses showed residents participated daily in *planned exercise activities* at a higher rate than either the social or outdoor activities. Seven hundred and-seventy seven residents (39.1%) participated in planned exercise activities daily, 334 residents (16.8%) participated less than daily, 221 residents (11.1%) participated about weekly, 135 residents (06.8%) less than once a week, 86 residents (04.3%) participated less than once a month and 435 residents or (21.9%) of the sample did not participate at

TABLE 1. Resident Participation in Planned Exercise, Social and Outdoor Activities

% Resident Participation in:	% Planned Exercise Activity	% Planned Social Activity	% Planned Outdoor Activity
Frequency			
Daily	39.1	30.8	04.6
Less than daily	16.8	20.7	05.3
About weekly	11.1	14.6	08.9
Less than once a week	6.8	11.2	16.8
Less than once a month	4.3	05.5	16.3
Not at all	21.9	17.2	48.1

Source: Staff interviews tailored to each of the 1,988 residents in the study. For less than 2% of subjects for each item, the staff member replied that they "did not know." Those responses are distributed across the categories in the same frequency as the actual.

all. For *planned social activities*, results for 1,988 residents showed that 612 residents (30.8%) participated daily, 412 residents (20.7%) participated less than daily, 290 residents (14.6%) participated weekly, 223 residents (11.2%) participated less than once a week, 109 residents (05.5%) participated less than once a month and 342 residents or (17.2%) did not participate at all in planned social activities. When asked about the amount the resident participated in *planned outdoor activities* the responses showed considerably less participation. Of the 1,988 resident responses, 956 residents (48.1%) did not participate in planned outdoor activities at all, 324 residents (16.3%) participated less that once a month, 334 residents (16.8%) participated less than once a week, 177 residents (08.9%) participated about weekly, 105 residents (05.3%) participated less than daily and only 92 residents (04.6%) participated on a daily basis. These numbers show very low involvement in outdoor activities by the residents in our sample.

Outdoor spaces per beds in the facility ranged from one outdoor space for 200 residents to three outdoor spaces for 55 residents. Table 2 shows the number of outdoor spaces by state, number of beds and number of units. New Jersey and California were tied for the most outdoor space, followed by Minnesota, New York, and Florida in last place. In this sample, climate did not determine the amount of outdoors spaces per bed, but the number of stories in the facility did.

We developed an Outdoor Amenities Index including 10 items with the potential to expand socialization and stimulation opportunities outdoors for residents, families and staff. Table 3 summarizes the percentage of 10 items (outdoor amenities) found at the unit level and 10 items (outdoor amenities) found at the facility level. Out of 131 units, only 58

TABLE 2. Summary of Number of Outdoor Spaces by State, Number of Beds and Number of Units

State Facility Located In	Number of Beds	Number of Units	Number of Facility Outdoor Spaces	Number of Unit Outdoor Spaces	Total Number of Outdoor Spaces
MN Total	1033	30	8	14	22
FL Total	1194	30	7	10	17
CA Total	818	22	8	11	19
NY Total	888	27	8	7	15
NJ Total	1052	22	8	16	24

TABLE 3. Outdoor Amenities at Unit and Facility Level

Outdoor Features at Unit Level (10 items) (n = 131units)	% of 131 nursing units with item	Outdoor features at Facility Level (10 items) (n = 40 facilities)	% of 40 facilities with item
Direct access to outdoor area from unit	44.3	Outdoor patio area	97.5
Outdoor seating	39.7	Flower garden	97.5
Outdoor table	35.9	Outdoor seating	95.0
Covered seating	33.6	Outdoor table	92.5
Covered patio area	33.6	Hard surface walking path at least 3'	87.5
Flower garden	33.6	Equipment for recreational activities	82.5
Outdoor area secured	33.6	Covered seating	82.5
Covered table	32.1	Covered patio area	82.5
Hard surface walking path at least 3'	26.0	Secured outdoor area	65.0
Raised garden planter	20.6	Raised garden planters	52.5

units (44.3%) had direct access to an outdoor environment. Thus, residents on 73 units (55.7%) had no access to the items featured on the outdoor amenities index without leaving their immediate unit. Even in those units with outdoor access, amenities varied between facilities and often access remained locked throughout the day. We did not ask each facility their policy on keeping the door to the outdoor area locked, although, during our assessment we noted whether residents were able to use the outdoor space independently without asking that the area be unlocked. In 6 units (10.3%) of the 58 units with direct access to the outdoors, there was no seating available, and when seating was available it was not covered 24.1% of the time. The outdoor space was secured or enclosed 75.9% of the time in the 58 units with direct access and an outdoor hard surface walking path at least 3 feet wide was available 58.6% of the time when direct access was available from the unit. (Areas were determined to be secure at unit or facility levels if they were enclosed either from the location of an inside courtyard or the area was fenced.)

Facility level amenities were assessed separately from those found at the unit level. Thirty-nine of the 40 facilities (97.5%) assessed had at least one outdoor space that included some amenities. Approximately two-thirds (65%) had a secured outdoor area, 87.5% had a hard surface walking path and 82.5% of the facilities provided a covered seating area. At this time we have not correlated resident usage of the outdoors with the level of amenities found.

We were also interested in the extent to which the facilities facilitated the ability of residents to watch activity or wait outdoors for transportation, or to sit comfortably immediately inside the facility, an issue in both cold winter climates and sunny humid climates. Very few facilities provided an outside area at the entrance to the facility where residents could sit and watch the activity of people coming and going or wait for transportation. When the space was available, it was a popular place with the residents, often preferred to an inner courtyard area.

Following the environmental assessment of each facility, descriptive narratives were written for each facility that described extraordinarily varied facilities and 131 varying units within them–some with an abundance of enrichments and some with substantial environmental deficits. Even an abundance of outdoor spaces did not guarantee that the spaces were supportive of residents using them independently. Generally, those facilities with limited outdoor spaces were large facilities both in the number of beds and number of stories high.

Providing outdoor spaces connected to special care units provide both a challenge and an opportunity. The challenge is to create a secure outdoor space that doesn't provide a view beyond the space because often residents will make an effort to elope beyond the parameters of the space. A very common mistake is to fence the area with a see-through material such as metal chain link. In one Florida SCU, a resident placed a chair on top of a table, climbed up on top of the chair and managed to climb over the chain link fence. No effort had been made to camouflage the fence with plantings or vines so the resident, who had lived in the area all of her life, recognized familiar landmarks beyond the fence and successfully left the confines of the outdoor space to join the world outside the fence. Unfortunately, the response of the administrator was to lock the access to this space rather than take steps to camouflage the fence.

In contrast, two excellent examples of outdoor spaces were found on SCU units, one in Florida and the other in Minnesota. An outdoor butterfly garden was created in a central courtyard area on the second floor between two nursing units, one SCU and one skilled, in a Florida facility. Automatic doors facilitated access to this enchanted space directly from both units. A simple code was required for those returning to the skilled unit while the door automatically opened for the resident returning to the SCU unit. Because there were walls on all four sides fencing was not necessary. Abundant plantings, flowers, trees, birdbaths, and fountains created an idyllic setting where butterflies flourished and families and residents visited.

A SCU unit in a rural Minnesota facility maximized the value of nature for the residents with good design of indoor and outdoor spaces. Large "windows to the world" with windowsills lined with African violets provided views of the ever-changing Minnesota weather along with vista views of the fields and river beyond. A secure outdoor patio was located directly off of the unit. Because this is a rural community and most of the residents lived on farms previous to coming to the home, an effort was made to create outdoor space that was familiar to the residents and space that was not just for passive enjoyment but space that could be used as an activity. A resident made good use of a push lawnmower cutting the grass on a daily basis and the ripe cucumbers and tomatoes became salads using recipes that somehow the residents had not forgotten. Each of the twenty residents on the unit had a sun hat that was conveniently located on the wall adjacent to the door leading to the patio. The patio area was just the perfect size for a small garden, walkways, patches of grass, a bright umbrella table and a glider where three friends sat together on a daily basis. Residents were encouraged to use this space independently.

Time and again, we assessed outdoor spaces that were plentiful in amenities but were not used by residents mainly because of the location of that outdoor space. A three-story New York facility, that was attached to a hospital, recognized the need for outdoor space and created a very attractive solarium-outdoor patio combination. This facility was unique in our sample in that it did not have a central reception area, but rather, administrative offices were located immediately inside the entrance. Offices, therapy rooms, a large activity room, a vigil room—co-occupied with staff computers—and the solarium/patio area were all located on the first floor with resident rooms located on the second and third floors. The new solarium room was located at the very end of the first floor. It is a lovely room with floor to ceiling windows and several skylights. Doors from the space lead directly to an outside area complete with lovely plantings and patio furniture. Staff was very proud of this new area and it is a lovely space but residents do not use it because of its distant proximity to resident rooms. The administrator admitted that the location is very problematic and is making efforts to better utilize the space for resident activities.

Unfortunately this scenario was repeated often in our assessment of the 40 facilities. Beautiful outdoor spaces were built, often with community support, but it was unrealistic to think that residents could make use of the spaces, either independently because of their distance from resident rooms or with the assistance of staff because of the time required to assist residents to the space. One facility built a lovely large chalet type screen porch 500 feet from the facility. Access was off the activity room, where previewing the residents was possible, but there was no hard surface path leading to the porch which greatly reduced accessibility for residents in wheel chairs. Staff at another county facility with offices on the first floor and resident rooms on the second and third floor was ecstatic when funding became available for the construction of an outdoor patio space. This facility was located in a rural area with a large amount of acreage available for the placement of the patio. Unfortunately, the choice was made to construct the patio directly adjacent to the parking lot. Efforts were made to camouflage the parked cars with plantings but soon those plantings died from the exposure to car exhaust. The patio was a great distance from the main entrance and no hard surface path led to the patio. Consequently, staff mainly utilized the space during their breaks.

Central patio areas often went unused for a variety of reasons. In some, access was locked, in others the space was small and close to resident room windows so residents felt as if they were invading the privacy of others, but mainly residents reported the space as "being boring." There were some exceptions to this such as the facility that enlisted a resident to open the door and greet all those that entered the outdoor space. The outdoor space became a place to socialize and to watch the rabbits in the "rabbit den."

In contrast, the most used outdoor spaces and indoor window spaces were those that either had a view of or were located in a setting of real life activity. Residents expressed how they enjoyed sitting outside the main entrance and watching the activity. For those residents who fear leaving the security of the building, views from windows provided an alternative. As part of a renovation, one facility designed a lounge adjacent to the entry with floor to ceiling windows. Chairs were strategically placed in front of the windows and residents waved to staff and visitors as they left and greeted those arriving. A special feature of the windows was that they overlooked a garden where rabbits frolicked on a regular basis. As far as usage of space, this lounge area was used constantly throughout the day.

An example of a view from a window that became an activity for residents, visitors, and staff was located in a facility in a small rural New York town. The people in the town had great school spirit and were active spectators of athletic functions. The facility was 5 stories high and had previously been a county independent living setting but had been transformed into a nursing home. The square building was built with single loaded corridors and huge windows; both in resident rooms and in each floor lounge. The facility was located across from the only high school in town. The perpetual movement of students and busses provided a daylong theater of sorts. The athletic fields were located on the school grounds providing the residents with a perfect view of all athletic events. From the top floors, it was such a good view that visitors often came during sporting events to watch the game with the residents. These are just a few of the examples where nature and outdoor activities were used to the advantage of those who preferred not to or simply could not physically utilize outdoors space.

The qualitative component of the assessment and the programmatic interviews yielded some positive examples on the way facilities promoted or discouraged getting residents out of doors. For example, one urban facility reached out to a community garden association and a youth group to create a model urban garden that could be used by residents and other neighbors on the cul-de-sac where the facility was located. Another facility utilized a fleet of donated cars to take residents for rides around campus. A third facility used golf carts to take residents on rides through the lovely residential area where the facility was located, the golf carts having the added advantage of offering fresh air and opportunity to interact with neighbors. In a facility characterized by rather crowded indoor spaces with largely 4-bedded rooms, the activities director expressed his philosophy favoring normal activity such as building projects and painting projects in the very pleasant outdoor space. This enabled residents to participate or watch as he made repairs and undertook practical projects of interest to many. That particular facility, not coincidentally, sported a scarecrow in its garden.

Discussion

This work has strong implications for further research. The approach positions us to study the amount of variation within facilities and the factors associated with that variation. A hierarchical cluster analysis has been done (with write-up underway) to determine whether we can classify facilities by the characteristics of their environments at the room, unit, and facility level and whether the physical environments are correlated with quality of life.

For this paper we have described physical environments where outdoor spaces and amenities were available but rarely used in contrast to those that were well-used, and places where windows to the world provided views of real life activities and those where little of interest was to be seen. Simply put, residents cannot use outdoor space if it is not available to them or the distance to traverse is too long for them to use outdoor space independently and staff is not available to assist them. This was often the case in multistoried buildings or when there is not direct access off of the nursing unit. Our impression is one of many missed opportunities–for example, places where plantings and bird feeders could have been established outside windows at the end of the corridors, and places where paved walkways or seating were needed.

Our study verified the paucity of opportunity for individual residents to be outdoors and their desire to do so. For almost 50% of the 1,988 residents in our sample, a staff member who knew them well said they never were included in outdoor programming. Furthermore, in the views of residents and of their family members interviewed for the quality of life study, close to 40% felt that they did not get outside as much as they wanted; these findings are even more compelling because of the propensity of nursing home residents and family members who refrain from criticisms and complaints in discussions of their satisfaction.

Certainly some facilities were located in areas where the outdoors was more difficult to utilize than others. A facility on an urban strip in San Jose comes to mind: fast food joints were located on either side; a six-lane road was located in front, and a crowded parking lot in back. A virtue could be made of the interest that would be generated by this commercial strip and programs could be developed to accompany residents outdoors, but the area was not conducive to safe outdoor life for the most disabled of the residents. In contrast were a number of facilities, including most of the county homes in the sample, where the location would permit numerous inviting spaces at relatively low costs but where the actual creation of safe outdoor places on the periphery of the buildings was minimal or nonexistent.

The checklist approach provided the bare bones of the availability of outdoor spaces, which could then be the building blocks for programmatic efforts. The actual extent to which and the way spaces are used depends on facility policies (including policies on permitting residents to be outside on their own), and facility practices such as having outdoor barbecues, encouraging family to go outside with residents on the grounds and making sure that seating and tables are clean, dry, and in good repair.

Other than fire egress regulations, federal regulations do not take into consideration outdoor spaces in the standards or the nursing home survey process. This seems rather peculiar based on the intent and goals of the federal regulations that apply to well-being of nursing home residents; one might expect encouragement of outdoor access or even minimal requirements for outdoor space in nursing facilities. When writing about the regulations, Karen Schoeneman (2004), a CMS employee involved in the survey process, noted:

> These regulations say that residents of nursing homes have the right to choices over their schedules, activities, and anything that is important to them (F242); that the environment should accommodate their needs and preferences (F246); that they can refuse treatment (F155); and that they should be helped by the home to attain their optimal quality of life (F240) and quality of care (F309). (p. 34)

The reality, however, is that even if spending time outdoors is a strong preference of an individual resident there are no regulations that require that outdoor space be provided by the nursing facility. At the state level, if regulations pertaining to the outdoor environment are in place, most often they apply to special care dementia units and emphasize the safety of the grounds.

In our view, outdoor space is an important part of the overall physical environment of any nursing home, and, therefore, outdoor spaces should be assessed for their availability, accessibility, functionality, aesthetic interest, and conduciveness to privacy and to social relationships. Our glimpse of 40 nursing homes suggested both promising approaches and that much room for improvement remains. We emphasize that viable outdoor space will depend not only on the physical environment itself but on a variety of programmatic and policy choices relating to the use of staff and volunteers and the resolution of the ever-present conflict between resident preferences and perceptions of responsibility for resident safety.

REFERENCES

Berry, L.L., Parker, D., Coile, R.C., Hamilton, D.K., O'Neill, D.D., & Sadler, B.L. (2004). Can better buildings improve care and increase your financial returns? *Frontiers of Health Services Management*, 21 (1).

Christenson, M. (1990). *Aging in the Designed Environment*. Binghamton, NY: Haworth Press.

Cooper-Marcus, C., & Barnes, M. (1999). *Healing Gardens: Therapeutic Benefits and Design Recommendations*. New York: John Wiley & Sons.

Cutler, L. (2000). Assessing the environment of older adults. In R. L. Kane & R. A. Kane (Eds.), *Assessing Older People: Measures, Meaning, and Practical Applications* (pp. 360-382). New York: Oxford University Press.

Day, K., Carreon, D., & Stump, C. (2000). The therapeutic design of environments for people with dementia: A review of the empirical research. *The Gerontologist, 40,* 397-416.

Denhardt, R.B., Denhardt, J.V., & Aristigueta, M.P. (2002) *Managing Human Behavior in Public and Nonprofit Organizations.* Thousand Oaks, CA: Safe Publications.

Eshelman, P.E., & Evans, G.W. (2002). Home again: Environmental predictors of place attachment and self-esteem for new retirement community residents. *Journal of Interior Design, 28*(1).

Gerlach-Spriggs, N., Kaufman, R., & Warner, Jr., S. (1998). *Restorative Gardens: The Healing Landscape.* New Haven, CT: Yale University Press.

Gold, M. F. (2004). Designs for extended living. *Provider,* November, 18-33.

Greene, A., Hawes, C., Wood, M., & Woodsong, C. (1998). How do family members define quality in assisted living facilities? *Generations,* 21(4): 34-36.

Healthfacts. (2005). *Vitamin D Deficiency: Common Cause of Many Ailments.* New York: Center for Medical Consumers.

Holick, M.F. & Jenkins, M. (2003). *The UV Advantage.* New York: Simon & Schuster.

Kahana, E. (1975) A congruence model of person environment interaction. In P.G. Windley, T. Byherts, & E.G. Ernst (Eds.), *Theoretical Development in Environments and Aging.* Washington, DC: Gerontological Society.

Kane, R. A. (2003). Definition, measurement, and correlates of quality of life in nursing homes: Towards a reasonable practice, research, and policy agenda. *The Gerontologist, 43*(2), 28-36.

Kane, R. A., Kling, K. C., Bershadsky, B., Kane, R. L., Giles, K., Degenholtz, H. B. et al. (2003). Quality of life measures for nursing home residents. *Journal of Gerontology: Medical Sciences, 58A*(3).

Kaplan, R. (2001). The nature of the view from home: Psychological benefits. *Environment & Behavior, 33*(4), 507-542.

Larson, J. & Kreitzer, M. J. (2004) Healing by design: Healing gardens and therapeutic landscapes. *Implications,* 10(2), 1-6.

Lawton, M. P. (1983). Environment and other determinants of well-being in older people. [Review] [49 refs]. *Gerontologist, 23*(4), 349-357.

Lawton, M. P., Brody, E. M., & Turner-Massey, P. (1978). The relationships of environmental factors to changes in well-being. *Gerontologist, 18*(2), 133-137.

Lawton, M. P., & Nahemow, L. (1973). Ecology and the aging process. In C. Eisdorfer & M. P. Lawton (Eds.), *Psychology of Adult Development and Aging* (pp. 619 674). Washington, DC: American Psychological Association.

Leibrock, C. (2000). *Design Details for Health.* New York: John Wiley and Sons.

Lindamood, S. (1979). *Housing, Society and Consumers.* St. Paul, MN: West Publishing.

Maslow, A. (1970). *Motivation and Personality.* New York: Harper & Row.

Preiser, Rabinowitz & White (1988). *Post-Occupancy Evaluation.* New York, NY: Nostrand Reinhold.

Rowles, G. D. (1978). *Prisoners of Space? Exploring the Geographical Experience of Older People.* Boulder, CO: Westview Press.

Schoeneman, K. (2004). Don't blame Obra! *Best Practices,* November/December, 34-36.

Ulrich, R. (1984). View through a window may influence recovery from surgery. *Science,* 224, 420-421.

Outdoor Environments
at Three Nursing Homes:
Focus Group Interviews with Staff

Anna Bengtsson
Gunilla Carlsson

SUMMARY. This study investigated how the outdoor environments at nursing homes for older persons were experienced and used to gain knowledge with implications for design. Focus group methodology was used to explore staff's view of how the residents experienced and used the outdoors. Two main themes and ten sub-themes were the result when the focus group interviews were analyzed. Theme one, *being comfortable in the outdoor environment,* describe the residents' special needs to be able to and dare to use the outdoors. The theme suggests a *precautionary design,* which promotes security and safety and protects from disturbance and negative impressions. The second main theme, *access to surrounding life,* describes the residents' needs for change and variety in

Anna Bengtsson is a PhD Student, and Landscape Architect, Department of Landscape Planning, Swedish University of Agricultural Sciences, Box 58, 230 53 Alnarp, Sweden (E-mail: anna.bengtsson@lpal.slu.se).

Gunilla Carlsson, PhD, Reg. OT, is affiliated with Department of Health Sciences, Division of Occupational Therapy, Lund University Sweden.

The Research Program Arts in Hospital and Care as Culture within Stockholm County Council has supported this work. The authors would like to thank all the participants in the study, and Caroline Hagerhall and Patrik Grahn for valuable help with data collection and comments.

[Haworth co-indexing entry note]: "Outdoor Environments at Three Nursing Homes: Focus Group Interviews with Staff." Bengtsson, Anna, and Gunilla Carlsson. Co-published simultaneously in *Journal of Housing for the Elderly* (The Haworth Press, Inc.) Vol. 19, No. 3/4, 2005, pp. 49-69; and: *The Role of the Outdoors in Residential Environments for Aging* (ed: Susan Rodiek, and Benyamin Schwarz) The Haworth Press, Inc., 2005, pp. 49-69. Single or multiple copies of this article are available for a fee from The Haworth Document Delivery Service [1-800-HAWORTH, 9:00 a.m. - 5:00 p.m. (EST). E-mail address: docdelivery@haworthpress.com].

Available online at http://www.haworthpress.com/web/JHE
© 2005 by The Haworth Press, Inc. All rights reserved.
doi:10.1300/J081v19n03_04

the everyday situation and suggests an *inspiring design,* which promotes
stimulation of senses and mind and provides positive impressions. *[Article copies available for a fee from The Haworth Document Delivery Service: 1-800-HAWORTH. E-mail address: <docdelivery@haworthpress.com> Website: <http://www.HaworthPress.com> © 2005 by The Haworth Press, Inc. All rights reserved.]*

KEYWORDS. Elderly, older, outdoor environment, health design, universal design, landscape planning

INTRODUCTION

Today nursing homes for older persons only accommodate the most
fragile persons and very few of them can enjoy the outdoor environment on their own, i.e., they are dependent on the caregivers and next
of kin to go outdoors. The positive effects of fresh air, daylight and experience and use of green outdoor environment are well documented
(Cohen-Mansfield & Werner, 1998; Küller & Küller, 1990; Küller &
Wetterberg, 1996; Rodiek, 2002; Ulrich & Parsons, 1992; Ulrich, 1984;
Ulrich, 1999) and the health effects arising from experience and use of
outdoor environments are greater the more weak and fragile a person is
(Ottosson & Grahn, 1998, 2005; Ulrich, 1999).

One factor influencing older persons' desires and possibilities to go
out is the experience of the usability of the environment, i.e., the possibilities to move around, be in and use the environment on equal terms
with other citizens (Iwarsson & Ståhl, 2003). The less competent a
person is, the greater is the impact of environmental demands on that
person (Lawton, 1968), which means that the existence of few environmental barriers supports people with low functional capacity to use the
environment. However, the use of the outdoor environment is also a
question of how well it provides the users with desirable experiences,
i.e., how attractive the environment is. The environment has to be restorative (Kaplan & Kaplan, 1989; Kaplan et al., 1998; Ulrich, 1999,
Ulrich, 2001) and instorative (Grahn, 2005; Stigsdotter & Grahn, 2003;
Stigsdotter & Grahn, 2002). A restorative environment gives opportunity for recovery, through the restorative experiences *Being away, Extent, Fascination* and *Compatibility* (Kaplan & Kaplan, 1989; Kaplan et
al., 1998), or as suggested in Ulrich's theory of supportive gardens
(Ulrich, 1999; Ulrich, 2001), a health care facility has the ability to improve health outcomes if it provides the restorative resources *Sense of*

control and access to privacy, Social support, Physical movement and exercise and *Access to nature and other positive distractions*. An instorative environment strengthens our identity and self-esteem and makes us feel part of a meaningful context. When the experiences and activities in the environment are in harmony with the user's background and character health, well-being and drive are promoted (Grahn, 2005; Stigsdotter & Grahn, 2003; Stigsdotter & Grahn, 2002). Whereas the concept of restorativeness focuses on the experiences in an environment that gives opportunity to recover and to improve health outcomes, the concept of instorativeness also adds the possibilities of gaining something more than recovery, something of existential value and reorientation in life that make us more fit to meet future misfortunes.

The knowledge of outdoor environment for older persons is diverse and comprises knowledge about environmental barriers and usability in the public environment (Carlsson, 2004), design issues and guidelines for elderly people's environment in general (Carstens, 1985; Stoneham & Thoday, 1996), but also for the more specific environment at nursing homes (McBride, 1999; Ousset et al., 1998). Several design-related studies of the outdoor environment at nursing homes focus on older persons with dementia (Beckwitt & Gilster, 1997; Cohen-Mansfield & Werner, 1999; Hoover, 1995; Zeisel & Tyson, 1999) and how to use horticulture as therapy (Jarrott & Gigliotti, 2004; Midden & Barnicle, 2004; Stein, 1997). Few studies particularly deal with the staff's perspective of the outdoor environment for older persons at nursing homes. Earlier studies on staffs' perspective have focused on older persons with dementia and mainly used quantitative techniques. For example, through questionnaire surveys to personnel, Cohen-Mansfield and Werner (1999) investigated characteristics and features of outdoor areas for people suffering from dementia, and Rappe and Lindén (2004) documented observations regarding indoor and outdoor plants in homes for people suffering from dementia. The understanding of how different environments are experienced and used is basic for creating attractive and usable environments. The staff's view is of particular importance as they obtain a comprehensive knowledge of the users' wishes, needs and capabilities in the environment. Many nursing homes accommodate people with diverse diagnoses, not only dementia, and therefore, it is important to elucidate staff's perspective under these circumstances.

The purpose of this study was to explore factors of importance for the use of the outdoor environment at nursing homes for older persons by asking staff about the residents' use and experience of the outdoors. The study focused on factors with implications for the design and content of the outdoor environment.

METHOD

Focus group methodology (Krueger, 1998) was used to explore the staff's view of how the residents experienced and used the outdoor environment at the three nursing homes. The focus group technique was considered the most suitable since it was the extent of the participants' views that was sought after and not their individual opinions (Morgan, 1998). A focus group discussion is evolutional in character, meaning that statements and opinions flow from one another (Krueger, 1998). Thus the interplay among the research participants gives rise to richer modulations.

Three cases were used primarily to gain richness and variety in data and to reveal salient points, rather than to lead to concrete comparisons. The data were collected as a part of a case study that aims at describing and triangulating aspects of outdoor environments at three nursing homes for older persons.

Settings

The cases were selected among nursing homes for older persons with access to their own outdoor areas in a city in the south of Sweden. In order to obtain information-rich cases and maximum variation the three cases were purposefully selected. The criteria for variation concerned differences in outdoor design and content, spatial relations (inside building/close surroundings/neighborhood) and location of facility. None of the cases was used systematically for therapy. The main characteristics of the cases are presented in Table 1. A detailed description is available in Bengtsson (2004).

Residents

The residents at the three nursing homes were aged 65 years and above and had physical and/or psychological disabilities that made them dependent on support from caregivers round the clock. Nevertheless, the nursing homes accommodated people with a wide range of functional capacity, from people with cognitive limitations due to dementia and people permanently in their beds due to physical limitations,

TABLE 1. Main Characteristics of the Three Cases Where the Focus Groups Were Conducted

	Case 1/Focus group 1	Case 2/Focus group 2	Case 3/Focus group 3
Outdoor areas	Large unfenced park, atrium, patios	Park with wire fence, balconies	Garden in part of courtyard fenced with wide planks, conservatory
Number of residents	88	31	24
Location of facility	Residential area, flats and detached houses	Residential area, detached houses, close to sea	Business, shopping and residential area, tower blocks

to the very few who were able to leave the nursing home without assistance using walking aids, wheelchairs, or even by bicycle.

Focus Group Participants

At each nursing home, four to five staff members participated in a focus group. To obtain information rich cases among the research participants, the superintendent at each nursing home was asked to recommend participants among the staff working closely with the residents and with experience of their use of the outdoors. Altogether 14 research participants were included, all of them women in the age span from 20 to 60 years. They had been working at the three health care facilities from just a few months up to about 20 years.

All research participants joined the study voluntarily and in reported results none could be identified by other situational or contextual factors.

Data Collection

The focus group interviews were conducted to give qualitative data through open-ended questions. One moderator and one assistant conducted one focus group interview at each of the three nursing homes in June and July, 2004. The moderator guided the discussion in the group and the assistant took notes and asked follow-up questions. The open-ended questions in the general interview guide (Patton, 2002) aimed at exploring and describing the staff's view of the outdoor environment for the residents concerning the experience and use of the outdoors from inside the building, in the close surroundings and in the neighborhood.

Each interview lasted approximately 90 minutes and all of them were tape-recorded. When half of the session had passed a plan of the building and outdoor environment was provided to the group to further support the discussion. The interviews were transcribed shortly afterwards.

Data Analysis

The raw data, i.e., transcriptions of focus group interviews, were systematized by the first author into different themes using the analytical approach *Meaning condensation* (Kvale, 1996; Giorgi, 1985). This approach is an empirical phenomenologically based method intended to (1) find natural meaning units in the interview texts, (2) explicate their main themes and (3) relate themes to the purpose of the study. To increase the credibility, the co-author who had not participated in the focus group interviews read the interviews and the authors discussed the themes until they reached agreement.

The main themes are presented with summary descriptions and illustrative quotes. Quotations were abridged and modified to clarify their content and then translated into English.

RESULTS

The analysis gave rise to two main themes and ten sub-themes. The first theme, *being comfortable in the outdoor environment*, is represented with four sub-themes: *sensitivity to weather, familiarity, security* and *calmness*. The second theme, *access to surrounding life*, consists of six sub-themes: *capacity for outdoor activity, sensual pleasures of nature, following the rhythm of life in nature, surroundings as a way to keep up to date, surroundings as a source to relate to past times* and *social potential of outdoor environments*.

Focus groups one, two and three in the results represent corresponding case numbers, described previously under Settings.

Being Comfortable in the Outdoor Environment

Sensitivity to Weather

The staff reported that the outdoors kept the residents healthy and made them happier, but the residents were sensitive to the weather. Particularly in focus groups one and two the staff discussed the weather, and as a woman in focus group one said: *"It is as if there is something wrong. No matter how nice it is or how much they see, if it is windy, then that's it."* Rain, wind, cold or snow were hindrances to going out since people felt cold and it could be slippery. Even if the staff pointed things out, the residents were not interested. *"They are totally occupied by being cold."*

The staff also discussed how the residents grumbled less when they were outdoors and they were content just to sit in the sunshine with their eyes shut. One of the staff expressed it in the following way: *"But just to sit in the sun and drink your coffee, it is obvious that everyone benefits from that."* Often the residents deliberated about whether to go out or not. *"If the weather is bad they are really sorry that they can't go out. The first thing in the morning, they ask: 'What is the weather like?' and 'Are we having coffee outdoors or indoors?'"* said one staff member.

Security

According to the staff it was important that there were sheltered and secure outdoor environments right beside the most frequently used common rooms in the building where the residents could easily get out by themselves. The staff in focus groups one and two discussed the importance of the proximity to the building and the staff. In focus group one, the staff said *"The patios feel secure because the windows and doors are usually open, they can hear sounds coming from indoors and the staff is close by."* In focus group three the staff described their own environment as close and secure and thought it was the reason for the frequent use when the weather was fine. They mentioned that the residents were close to one another and the staff in the garden and that the fence of wide planks prevented outsiders from coming too close to the residents. They thought that this gave a feeling of security and that a larger garden might have been less pleasurable for the residents (Photo 1). The staff in focus group one mentioned that people with visible disabilities might feel uncomfortable when being viewed by outsiders. Therefore they needed places that prevented them from being looked at. Sometimes this was expressed by their next of kin and not by the residents themselves. The staff in focus group one considered the patios at their nursing home as fairly shut off from people's view.

The three focus groups considered it important that the outdoor environment was easy and secure to walk in. For example, in focus group one the participants said that the railing was a great source of security

PHOTO 1. The Small Enclosed Garden in Part of a Courtyard Discussed in Focus Group Three

for the residents. Other aspects increasing the security were low thresholds and edges, and that the lifts were easy to use. However, one of the staff mentioned that the environment should not be too perfectly designed and doctored but rather informal.

Focus groups two and three also discussed the staff's own insecurity about letting the residents be out on their own. Focus group two mentioned the risk of someone suffering from dementia wandering off if there was insufficient supervision. In focus group three a worry was expressed that people could fall outside on stairs or into the pond, especially during the winter season when it was slippery and the plants around the pond had not yet grown.

Familiarity

According to the staff, people suffering from dementia, in particular, needed a familiar environment with people around them that they recognized. A feeling of unfamiliarity could cause great problems. In focus groups two and three the staff recognized the nursing home and the park/garden as familiar environments. The residents felt at home and this was safe for them. In focus group one the patios were the safe out-door environments and to a varying degree the surrounding park and neighborhood were used on the residents' terms. Staff and residents did not need to go far from the building for most of the residents to experience that they were somewhere else. A distance perceived as very short by the staff could be perceived as quite far by many of the residents (see Photo 2). For most of those that did enjoy getting away from the nursing home, one hour was sufficient.

Calmness

Focus groups two and three discussed calmness as a particularly im-portant quality in the environment, especially with regard to people suffering from dementia. The staff considered calmness essential to pre-

PHOTO 2. The Large Unfenced Park in a Residential Area Discussed in Focus Group One

vent restlessness. Too many people counteracted the quality of calmness. Focus group two agreed about the calming influence of the sea (Photo 3) and focus group three considered the garden in itself to be calming (Photo 1). Focus group three also mentioned the fountain as a certain source of calmness and relaxation. A lot of residents therefore chose to sit next to it.

Access to Surrounding Life

Capacity for Outdoor Activity

Statements of the various ways that the outdoors was used recurred in the three focus groups. The span ranged from those who never left their room to those who went out in all weathers and during all seasons. These differences depended upon ability as well as personality according to the staff.

For most of the residents a little was enough, and they used the park or garden for short walks, sitting and relaxing, sitting and talking, drink-

PHOTO 3. The Open Park Close to the Sea Discussed in Focus Group Two

ing coffee, looking at the surroundings or reading the newspaper. In focus group one the staff made a few concrete comments about how the residents' use of the outdoors was limited when the greenery was not maintained adequately. For example, it was difficult for the residents sitting in the patios to see over the bushes and it was difficult to use the lawn areas when the grass was not properly mown.

The staff discussed that there should not be too much at a time and that one activity was enough for one day for many residents. *"For example on Fridays, Bengt visits and reads aloud from books on old Malmö and talks to everybody. After that it doesn't fit to ask: 'Shall we go outside, Marianne?'"* Entertainment in the garden, such as a midsummer celebration or a barbecue, was very well appreciated among some of the residents, while others, in particular people suffering from dementia, could be very worried. A few residents wanted to be outside all day. *"They only have time to go indoors to go to the bathroom or take their meals. And they sit and get a sun tan."* *"Well, I'm going outdoors again, they say."*

The staff considered the surroundings outside their own environment important. They accentuated the importance of being able to choose: *"If I feel like walking around the houses, that's what I do. If I feel like watching the sea, that's what I do. If I feel like sitting in the garden, that's what I do."* Assisted by staff or next of kin, some of the residents enjoyed going away for longer trips or getting into the city and seeing the crowds in the street. A few residents left the nursing home area on their own and could stay away for several hours.

Sensual Pleasures of Nature

The significance of contact between the residents and natural elements was noticeable when the staff discussed the value of bringing in plants, leaves, flowers and branches, and placing them on the tables. Smelling, feeling and discussing the flowers was a source of great joy for the residents. Fruit and berries were appreciated, just to taste or to bake a cake with them. Further they mentioned the importance of getting daylight into the rooms and that the view from the window should provide color, flowers and greenery. In focus group one residents and next of kin had been very upset when the plants were allowed to grow over the windows for a long time and it got dark and confined in the rooms.

According to the staff, it was important for the residents to go outdoors to sense the freedom. *"You do not feel confined in a garden. When you come to the garden you feel somewhat more alive."* They mentioned that outdoors you can feel the wind against your skin, the scent of flowers or new-mown grass, you can take off your shoes and feel the grass against your feet and you get fresh air and daylight. Some of the residents only wanted to go out to enjoy the sun and the heat of the sunshine. *"O, look how tanned I have become,"* the residents would say. One staff member emphatically talked about the lady who took so much pleasure in being out in the summer rain, to hear the drops and feel the splash against her skin. Another told of a walk with one of the residents: *"When she saw the lilac we had to cross the street and take the wheelchair right up to the bush for her to really smell the fragrance. It was very important to her. To feel the flowers, to somehow come close."* Contact with animals was also mentioned in the focus groups. The staff recalled that the residents made comments about the birds outside in wintertime and that they often saved breadcrumbs to feed the birds. Everyone appreciated when visitors brought their dogs.

Following the Rhythm of Life in Nature

The staff mentioned that the vegetation was important to be able to follow the seasons. When spring came the residents talked about it starting to sprout and they followed the progress of the flower buds. They noticed new seasons and periods through different flowers. *"It is the same for them as it is for us, when winter arrives you creep into your nest and when the sun comes you go back out again,"* one staff member said. Midsummer was a real highlight celebrated outdoors at the three nursing homes. According to tradition they picked flowers for the maypole and there were music and dance performances in the garden. The color of autumn, the variation when the trees turned red and orange and the arrival of horse chestnuts was also much enjoyed by the residents according to staff. Although the staff spent most time talking about experiences of other seasons than winter, they mentioned how beautiful and different it was with snow and white frost during winter. Focus group one mentioned that the staff usually brought Christmas trees and Christmas lights to the outdoor environment for the residents to see the light glimmer through the windows.

The importance of being able to follow the weather and seasons in the outdoor environment was also noticed by the staff through residents' comments such as: *"Oh, the sun is shining!"* or *"Oh, it's pouring out-*

side!" When indoors the residents often talked about the weather and looked at the trees to see if it was windy outside.

Surroundings as a Way to Keep Up to Date

The staff noticed that contact with people and society outside the nursing home was important for the residents. Residents talked about what was happening outside, people and dogs walking past or cars and bicycles passing by. According to the staff the residents were happy to see something other than what happened in the nursing home, and they appreciated that there was life going on around them. In focus groups one and three the staff mentioned the importance of the seating places in the entrances where people came and went and a lot was happening. The most alert of the residents tended to sit there. The staff considered it positive that they tried to get involved with life around them to keep up their intellect and interest.

In focus groups one and two the outdoor environment had visual access to the neighborhood, which was perceived as beneficial for the residents. In focus group three there was no visual access from inside the garden (Photo 1). but this was not perceived as a disadvantage by the staff since the contact with the outside was provided from the inside the building.

The staff recalled how important it was for the residents not to be divorced from reality but to change environments every now and then, to get to see what it was like outside and to meet other people. One staff member talked about a lady at the nursing home with whom she went out every week: *"She doesn't buy that much but she likes just going into the shops. She sits in a wheelchair. Just to see people and to see what's in the shop, to see what has changed. And I can understand that that is very important for her."* In focus group three it was pointed out that it was an advantage to live centrally because it was just a short walk from facilities like hairdressers, foot care, the hospital, the pharmacy and shops. The residents enjoyed looking in shop windows even if they were not going to buy anything. Instead residents and staff could laugh together at the changing fashions.

Surroundings as a Source to Relate to Past Times

The staff explained how the outdoor environment wakened memories and helped the residents to relate to their life and their past. *"Look at that hydrangea, I had one when I . . ."* Plants and elements outdoors made the residents talk about their gardens, what they looked like, what flowers they had and what flowers lasted well. Also the surrounding environment could bring up habits from the past. In focus group one the staff talked about a man that had his own garden and also worked with gardens and plantations in his past: *"He came to me this spring and asked when we were going to sweep the patios and prepare them, because we did that the year before. I mean the memories are there somehow. I was a bit surprised, but we did it and he liked it. He used to have a garden and a house and he is used to sweeping and pottering about."* The surroundings could also lead to disappointment, as for this man who had worked with gardening and knew how it should be maintained. He was very disapproving when it was neglected. In focus group three staff mentioned the walk around the church as much appreciated. The residents then talked about when they had been to church for weddings and christenings.

Social Potential of Outdoor Environments

The staff explained how social intercourse was different outdoors. The fellowship between different residents and between residents and visitors was increased outdoors. Outdoors seats were free and everybody was mixed, compared to indoors where people tended to guard their own areas and keep to their own places. The garden was thus considered more flexible by staff. If staff was sitting with the residents and a visitor arrived it was easy to provide a place for them. The garden was considered a venue for meetings. Staff mentioned that whereas the residents never visited each other's rooms, they met and mixed with people from other floor levels in the gardens. It was an event for them when they would meet again after a few days and could exchange some news.

One staff explained that it was different to talk outside: *"It is easier to say: 'What kind of a flower is that? What's the name of it?' If you don't know someone else knows and there is a discussion about this flower. Somehow it is easier when you have something to focus the conversation on. That is more easy when outdoors."* According to the staff, they and the residents discussed whatever they saw around them: plants, houses, cars, people and dogs, for instance. In focus group two they mentioned how they made jokes about the smell of seaweed, which was appreciated by some and disliked by others. Also they made jokes about who was living in the small pavilion in the garden (Photo 3).

The staff also explained how a walk with a resident gave them a certain time together. This meant that they could attend to one resident only and talk as long as they wanted to. Also for people suffering from dementia the interaction with others could be improved outdoors. One staff member explained: *"It is better for them if you go out with one at the time, then they get undivided attention and they can attach to one and the same. It is easier for them to focus their minds. It is more confusing to be indoors."* Visitors were often more cautious indoors and they were easily troubled by other residents, the staff said. Outdoors they could speak freely. *"You can gossip about the staff if you like,"* one staff member added.

Another aspect of the outdoors mentioned by staff concerned the possibility to be proud of being in an attractive place that people admired and enjoyed. The staff proudly told how visitors, when coming for the first time, admired the view and the park or garden. They were then happy to say that they could use it as much as they wanted. Staff also mentioned how much better it must feel for the next of kin to the residents that the surroundings of the nursing home were beautiful.

DISCUSSION

A usable and attractive environment increases the possibilities and the drive to go outdoors both from the staff's point of view and from the residents' according to this study. Earlier studies show that if the close outdoor environment is well designed, this increases the number of visits to other recreational areas. This is valid for the population in general (Grahn & Stigsdotter, 2003) and at nursing homes for older persons (Grahn, 1988). This study aimed at describing factors of importance for the use of the outdoor environment at nursing homes for older persons, factors with implications for the design and content of the outdoor environment.

This study did not build upon the presumption that one outdoor environment could be ideal for all nursing homes for older persons. Rather it aimed at an increased understanding of the variety of roles that the outdoor environment plays and how this affects experience and use. The study did not intend to generalize but to describe the depth and complexity of the topic. The understanding that the results gave rise to is intended to be converted (by its readers) to other real life situations as regards use and design of outdoor environments for old and fragile people in need of care. This qualitative way of generalization is described by Guba and Lincoln (1989) as transferability and by Stake (1995) as naturalistic generalization.

The investigation gave rise to two main themes. The first theme, *being comfortable in the outdoor environment,* particularly connects with aspects of the residents' special needs to be able to and dare to use the environment, and the second theme, *access to surrounding life,* describes the outdoor environment as a source of change and variety in the everyday situation, to receive stimulation and to get away and feel free. The two themes counterbalance one another. On one hand, the residents were reported to be very sensitive and needed a design that promoted security and safety and protected from disturbance and negative impressions, i.e., *precautionary design,* and on the other hand a design that promoted stimulation of senses and mind, i.e., an *inspiring design.* It is important not only to design with the limitations of the users in mind but to provide opportunities and stimulation for the most fragile as well as for the more healthy of the residents. This is described as *compatibility,* the fit between one's inclinations and environmental circumstances by the Kaplans (Kaplan & Kaplan, 1989; Kaplan et al., 1998). Also, to create an instorative environment, i.e., an environment that recognizes our personality, it is important to stimulate abilities and give possibilities for development.

The duality expressed by the two main themes relates to the concept of usability by describing and exemplifying basic needs and restrictions for a person to *be able to move around, be in and use the outdoor environment.* Also, a person should be able to use the environment *on equal terms with other citizens,* which not only suggests that one has the possibility to go out whenever one wants to but also to get stimulation and opportunities for a varying degree of passive and active activities depending on personal wishes.

The first theme, *being comfortable in the outdoor environment,* gives implications for the *precautionary design.* Here the immediate surroundings are of the highest priority since a lot of the residents only get this far. The close environment is secure, easily reached and provides elements that help the residents to feel at home and easily recognize the surrounding. The *precautionary design* includes places protected from wind and rain, places that cannot be viewed by outsiders and places where you can be on your own. Further the design is barrier-free and avoids doctored design solutions that clash with the homelike quality of the environment.

The *precautionary design* connects to the restorative resource *Control* suggested by Ulrich (1999) in the theory of supportive gardens. The resource *Control* is described as the ability of persons to affect their situation and determine what to do and what others might do to them. This implies an environment that supports the user to the highest possible degree. Control is especially important to enforce under circumstances where the caregivers possibly support rather too much than too little. The results of this study indicate that a *precautionary design* could make the staff allow the residents to use the outdoor environment more on their own. Being able to and daring to use the outdoor on one's own also connects to the concept of instorativeness, which suggests that the ability to use the outdoor environment can restore a person to a more positive view of himself and his capacities.

The second theme, *access to surrounding life*, speaks for another side of the design, the *inspiring design* that promotes stimulation of senses and mind. This design provides views of colorful plants and traditional elements that connect with the seasons and the residents' earlier life, not only when using the outdoors but also from inside the building. It gives opportunities to see people coming and going to the facility or persons, bicycles, dogs, etc., passing by. The design gives the possibility to come close to sweet-smelling vegetation, to touch leaves and branches, to hear purling water and other things that stimulate the senses. It is a flexible design that gives possibilities to meet and to socialize in large as well as in small groups.

The inspiring design provides the users with restorative experiences when things and events in the surroundings catch the involuntary attention, i.e., *fascination*, and by creating opportunities to get away from pressures and obligations and thus restore directed attention, i.e., *being away* (Kaplan & Kaplan, 1989; Kaplan et al., 1998). Further, the inspiring design generates the restorative resource *natural distractions* that gives opportunities to distance oneself from negative situations, i.e., *temporary escape*, which is fundamental to the restorative resource *control* (Ulrich, 1999). An inspiring design give the user opportunity to take part in real life and be part of the world of meaning, which is essential to the instorativeness of the environment (Stigsdotter & Grahn, 2003).

The importance of an inspiring design that relates to the earlier life of the residents in particular corroborates earlier studies that show the importance of an environment that confirms one's self-concept. For example, Grahn (1991) found that when nursing home residents used green areas, it was most important that the surroundings were reminiscent of environments in which they were at their most active. Küller (1991) found that routines and environments that were more homelike than hospital-like helped the residents to maintain a healthier and less confused state.

The inspiring design that gives opportunities to meet with other residents, visitors and staff as well as people from outside the facility, connects to the restorative resource *Social support* in the theory of supportive gardens. Social support is motivated by the fact that there is a general association between the number of social contacts and health status (Ulrich, 1999). This statement is interesting to put into relation to *the model of involvement* (Stigsdotter & Grahn, 2002). This model suggests that the stronger the mental capacity of a person, the more involvement with other people, and the opposite: the weaker the mental capacity of a person, the less the involvement with other people. Whereas contact with others is considered health-providing in the theory of supportive gardens, the model of involvement suggests that contact with others depends on a person's health status. Either way the design must provide social interaction in a flexible way so that people have the possibility to spend time together in groups of varying size, and also to spend time alone.

This study assumed the staff's insights as particularly important for understanding how the outdoor environment is experienced and used by the residents. Their statements originate not from one single individual's particular needs but from an overall picture of many different residents. The focus group interviews proved their understanding and empathy for the residents and thus the results give implications for design for the resident's support and benefit rather than for the staff's. However, the results need to be related to the residents' as well as next of kin's views to further develop the two concepts of *precautionary design* and *inspiring design,* to promote design that counteracts a situation that focuses only on necessary measures and instead creates opportunities to fully live out one's whole life.

REFERENCES

Beckwith, M. & Gilster, S. (1996). The paradise garden: A model for designing for those with dementia and Alzheimer's disease. *Journal of Therapeutic Horticulture,* 13, 45-52.

Bengtsson, A. (2004). Staff's views on outdoor environments for elderly people–Focus group interviews at three nursing homes. In conference proceedings *Open Space People Space*: http://www.openspace.eca.ac.uk/conference/proceedings/start.htm.

Carlsson, G. (2004). Travelling by urban public transport: Exploration of usability problems in a travel chain perspective. *Scandinavian Journal of Occupational Therapy,* 11, 78-89.

Carstens, D. (1985). *Site Planning and Design for the Elderly: Issues, Guidelines, and Alternatives*. New York: Van Nostrand Reinhold.

Cohen-Mansfield, J. & Werner, P. (1998). Visits to an outdoor garden: Impact on behaviour and mood of nursing home residents who pace. In B. Vellas, J. Fitten and G. Frisoni (Eds.), *Research and Practice in Alzheimer's Disease* (pp. 419-436). New York: Springer Publishing Company.

Cohen-Mansfield, J. & Werner, P. (1999). Outdoor wandering parks for persons with dementia: A survey of characteristics and use. *Alzheimer Disease and Associated Disorders*, 13(2), 109-117.

Giorgi, Amadeo (1985). *Phenomenology and Psychological Research*. Pittsburgh, PA: Duquesne University Press.

Grahn, P. (1988). *Egen härd-guld värd: Institutioners och föreningars behov av egna grönområden* [*There is no place like home: On institutions' and organizations' needs for their own green areas*. In Swedish], Stencil 88:8, Institutionen för landskapsplanering, SLU, Alnarp.

Grahn, P. (1991). *Om parkers betydelse [On the significance of parks*. In Swedish], Stad & Land 93, Institutionen för landskapsplanering, SLU, Alnarp.

Grahn, P. (2005). Att bota sjuka i en trädgård-om trädgårdsterapi och terapeutiska trädgårdar. In M. Johansson & M. Küller (Eds.), *Svensk Miljöpsykologi [Swedish Environmental Psychology*. In Swedish], Lund: Studentlitteratur, in press.

Grahn, P. & Stigsdotter, U. (2003) Landscape planning and stress. *Urban Forestry & Urban Greening*, 2, 1-18.

Guba, E. & Lincoln, Y. (1989). *Fourth Generation Evaluation*. Newbury Park, CA: Sage.

Gurski, C. (2004). Horticultural therapy for institutionalized older adults and persons with Alzheimer's disease and other dementias: A study and practice. *Journal of Therapeutic Horticulture*, 15, 25-31.

Hoover, R. (1995). Healing gardens and Alzheimer's disease. *The American Journal of Alzheimers Disease*, March/April, 1-9.

Iwarsson, S. & Ståhl, A. (2003). Accessibility, usability and universal design-positioning and definition of concepts describing person-environment relationships. *Disability and Rehabilitation*, 25, 57-66.

Jarrott, S. & Gigliotti, C. (2004). From the garden to the table: Evaluation of a dementia specific HT program. In D. Relf (Ed.), *Expanding Roles for Horticulture in Improving Human Well-Being and Life Quality*. Acta Horticulturae 639, 139-144.

Kaplan, R. & Kaplan, S. (1989). *The Experience of Nature: A Psychological Perspective*. New York: Cambridge University Press.

Kaplan, R., Kaplan, S. & Ryan, R. (1998). *With People in Mind: Design and Management of Everyday Nature*. Island Press, Washington, DC.

Krueger, R. (1998). *Analyzing & Reporting Focus Group Results*. Thousand Oaks, CA: Sage Publications.

Küller, R. (1991). Familiar design helps dementia patients cope. In W. F. E. Preiser, J. C. Visher & E. T. White (Eds.), *Design Intervention: Toward a More Human Architecture*. New York: Van Nostrand Reinhold, 255-267.

Küller, R. & Küller, M. (1990). Health and outdoor environment for the elderly. In H. Pamir, V. Imamoglu, N. Teymur (Eds.), *Culture-Space-History*, Proceedings 11th International Conference of the IAPS.

Küller, R. & Wetterberg, L. (1996). The subterranean work environment: Impact on well-being and health. *Environment International*, 22, 33-52.

Kvale, S. (1996). *Interviews: An Introduction to Qualitative Research Interviewing.* Thousand Oaks, CA: Sage Publications.

Lawton, P. & Simon, B. (1968). The ecology of social relationships in housing for the elderly. *The Gerontologist*, 8, 108-115.

McBride, D. (1999), Nursing home gardens. In C. Cooper Marcus, and M. Barnes (Eds.), *Healing Gardens*. New York: John Wiley & Sons, 385-436.

Midden, K. & Barnicle, T. (2004). Evaluating the effects of a horticulture program on the psychological well-being of older persons in a long-term care facility. In D. Relf (Ed.), *Expanding Roles for Horticulture in Improving Human Well-Being and Life Quality*. Acta Horticulturae 639, 167-170.

Morgan, D. (1998). *The Focus Group Guidebook*. Thousand Oaks, CA: Sage Publications.

Ottosson, J. & Grahn, P. (1998). *Utemiljöns betydelse för äldre med stort vårdbehov. [The significance of outdoor environment for older persons in need of institutional care.* In Swedish], Stad & Land, 155:1998, Alnarp.

Ottosson, J. & Grahn, P. (2005). A comparison of leisure time spent in a garden with leisure time spent indoors on measures of restoration in residents in geriatric care. *Landscape Research*, 30(1), 23-55.

Ousset, P., Nourashemi, F., Albarede, J. & Vellas, P. (1998). Therapeutic gardens. *Archives of Gerontology and Geriatrics*, suppl. 6, 369-372.

Patton, M. Q. (2002). *Qualitative Research & Evaluation Methods*. 3rd ed. Thousand Oaks, CA: Sage Publications.

Rappe, E. & Lindén, L. (2004). Plants in health care environments: Experiences of the nursing personnel in homes for people with dementia. In D. Relf (Ed.), *Expanding Roles for Horticulture in Improving Human Well-Being and Life Quality*. Acta Horticulturae 639, 75-81.

Rodiek, S. (2002). Influence of an outdoor garden on mood and stress in older persons. *Journal of Therapeutic Horticulture*, 13, 13-21.

Stake, R. (1995). *The Art of Case Study Design*. Thousand Oaks, CA: Sage Publications.

Stein, L. (1997). Horticultural therapy in residential long-term care: Applications from research on health, aging, and institutional life. In S. Wells (Ed.), *Horticultural Therapy and the Older Adult Population*. New York: Haworth Press.

Stigsdotter, U. & Grahn, P. (2002). What makes a garden a healing garden? *Journal of Therapeutic Horticulture*, 13, 60-69.

Stigsdotter, U. & Grahn, P. (2003). Experiencing a garden: A healing garden for people suffering from burnout diseases. *Journal of Therapeutic Horticulture*, 14, 38-49.

Stoneham, J. & Thoday, P. (1996), *Landscape Design for Elderly and Disabled People*. Woodbridge: Garden Art Press.

Ulrich, R. (1984), View from a window may influence recovery from surgery. *Science*, 224, 420-421.

Ulrich, R. (1999). Effects of gardens in health outcomes: Theory and research. In C. Cooper Marcus and M. Barnes (Eds.), *Healing Gardens* (pp. 27-86). New York: John Wiley & Sons.

Ulrich, R. (2001). Effects of healthcare environmental design on medical outcomes. In A. Dilani (Ed.), *Design & Health: The Therapeutic Benefits of Design*. Stockholm: Svensk byggtjänst.

Ulrich, R. & Parson, R. (1992). Influences of passive experiences with plants on individual well-being and health. In D. Relf, *The Role of Horticulture in Human Wellbeing and Social Development* (pp. 93-105). Portland: Timber Press.

Zeisel, J. & Tyson, M. (1999). Alzheimer's treatment gardens. In C. Cooper Marcus and M. Barnes (Eds.), *Healing Gardens* (pp. 437-504). New York: John Wiley & Sons.

The Role of Design in Inhibiting
or Promoting Use of Common Open Space:
The Case of Redwood Gardens, Berkeley, CA

Galen Cranz
Charlene Young

SUMMARY. Redwood Gardens is a housing project for the elderly on the Clark Kerr campus affiliated with the University of California at Berkeley. A post occupancy evaluation study done in 1994 indicated that outdoor spaces were not well used, and a follow-up set of interviews and observations over a decade later in 2005 shows that the spaces are still underutilized, despite being greatly appreciated as an amenity. This paper explores reasons for the under usage, describes the design of the spaces, and makes proposals for improvements in seating and other environmental props that might facilitate both solo and group use of outdoor spaces. Design and policy recommendations are directed both to this immediate site and other such housing sites elsewhere. *[Article copies available for a fee from The Haworth Document Delivery Service: 1-800-HAWORTH. E-mail address: <docdelivery@haworthpress.com> Website: <http://www.HaworthPress.com> © 2005 by The Haworth Press, Inc. All rights reserved.]*

KEYWORDS. Outdoor space design, elderly housing, niches

Galen Cranz, PhD, is Professor of Architecture with a PhD in sociology, author of *The Politics of Park Design*, and recipient of the EDRA (Environmental Design Research Association) Achievement Award for 2004 for *The Chair* (E-mail: granz@berkelely.edu).

Charlene Young, B.Arch, is a 2004 graduate of the Department of Architecture, University of California, Berkeley, CA and is currently working in affordable housing with Devine & Gong, Inc., Oakland, CA.

[Haworth co-indexing entry note]: "The Role of Design in Inhibiting or Promoting Use of Common Open Space: The Case of Redwood Gardens, Berkeley, CA." Cranz, Galen, and Charlene Young. Co-published simultaneously in *Journal of Housing for the Elderly* (The Haworth Press, Inc,) Vol 19, No. 3/4, 2005, pp. 71-93; and: *The Role of the Outdoors in Residential Environments for Aging* (ed: Susan Rodiek, and Benyamin Schwarz) The Haworth Press, Inc., 2005, pp. 71-93. Single or multiple copies of this article are available for a fee from The Haworth Document Delivery Service [1-800-HAWORTH, 9:00 a.m. - 5:00 p.m. (EST). E-mail address: docdelivery@haworthpress.com].

Available online at http://www.haworthpress.com/web/JHE
© 2005 by The Haworth Press, Inc. All rights reserved.
doi:10.1300/J081v19n03_05

FITNESS THROUGH DESIGN?

The architect of Redwood Gardens on the Clark Kerr campus of the University of California at Berkeley, Sandy Hirshen, took pride in designing this low-cost HUD-sponsored housing for the elderly and physically disabled with special amenities (Figure 1). In 1994, eight

FIGURE 1. This postcard view was issued when Redwood Gardens opened in 1986.

years after its opening in 1986, about 80 undergraduates in the Department of Architecture under the direction of the senior author conducted a post occupancy evaluation (POE) of this housing complex. The building was evaluated with regard to residents' perception of security, their interaction with students, their use of university facilities, and their use of special design features in the interior and exterior of the complex. One of the special features, an enclosed, secure courtyard between two buildings with an amphitheater and allotment gardens, was verbally appreciated but underused (Cranz, 1994) (see Figure 2).

FIGURE 2. Site Plan

Redwood Gardens Site Plan

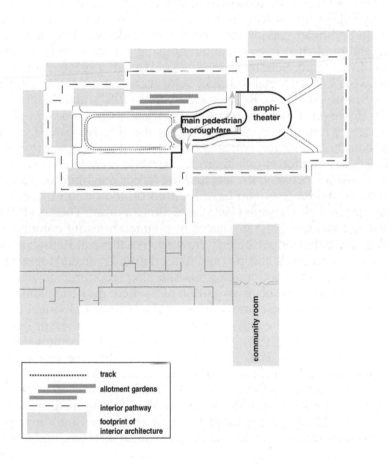

Today, this is more than a mild disappointment because policy-makers are explicitly worried about a lack of exercise in the American population as a whole, including its seniors. The Surgeon General's 2005 report blames the modern sedentary lifestyle for these problems (U.S. Department of Health and Human Services, 2001). We emphasize that the report calls for providing safe and accessible recreational facilities for residents of *all ages*. In 2001, Surgeon General David Satcher noted that health issues are not just a personal matter, because the way communities are organized affects individual health (U.S. Department of Health and Human Services, 2001). The goal of creating physical fitness through daily activity provides a new way, and another reason, to evaluate the use and design of these spaces. Accordingly, we revisited this site in 2005 in order to see if the courtyard was still underused, and if so, to investigate in what ways its design might have hindered its use.

Policymakers are looking to the design of outdoor spaces to see if they help or hinder routine exercise for the elderly. Many have conducted and published research about housing for the elderly: Pastalan (1975), Cranz and Schumacher (1975), Howell (1980), Cranz and Christensen (1978), Hogland (1985), Schwarz and Brent (1999), Regnier (1994), Brummett (1994), Regnier and Pynoos (1987), and Regnier et al. (1995). However, few have focused specifically on the design of outdoor spaces within the sites of housing for the elderly. Instead, discussion of common space has typically been defined as the space inside the building, and attention to neighborhood issues is most likely to concern the relationships between the site and the larger neighborhood and urban context. Those few who have focused on the design of common open spaces include Carstens (1985), and Stoneham and Thoday (1994).

Carsten considers outdoor spaces in planned housing communities for relatively independent, elderly citizens, like those at Redwood Gardens. She recommends gazebos and other naturally shaded meeting areas where residents may congregate. She also recommends that each person should have his/her own garden plot where they can plant whatever they choose. Some communities unwisely limit individual choice for the sake of creating a uniform landscape–at the expense of individual expression of creativity, which Carsten assumes people value, including both residents and managers. Additionally, she recommends raised planting beds for those confined to wheelchairs.

The British study, by Stoneham and Thoday, focuses on how to design public open space in order to secure the lifelong pleasures of the outdoors for elderly people in all kinds of residential building types. They consider access, seating, planting, wildlife interests, and more.

The recent concern for physical fitness puts a new perspective on earlier questions about the use and design of shared outdoor spaces. Accordingly, this paper makes a modest addition to the knowledge that architects, planners, facility managers, and gerontologists have about planning in designing common open spaces in congregate housing for the elderly.

METHOD: COMPARISONS OVER TIME, 1994 AND 2005

Returning to Redwood Gardens in 2005 offered the opportunity to see if this particular housing project for the elderly had changed in its patterns of usage of outdoor spaces compared to a decade earlier. Accordingly, the two authors conducted a week of observation (direct and photographic) and interviewing (both face-to-face and by telephone) between 5 and 12 March 2005. (For 2 weeks in the spring of 1994, 80 student researchers grouped in 20 teams of 4 had used similar methods of data collection: interviews, questionnaires, behavior traces, and behavior of observation.) Our 2005 visits were in the daytime, either late morning or afternoon. We spoke informally to those who happened to be outdoors, both residents and staff, for a total of 10 face-to-face on-site interviews. The interviews varied in length; three short interviews, with elderly residents, were 5-10 minutes, and four long interviews, with staff and with one of the few younger disabled residents, were 40-60 minutes. Several were repeat follow-up interviews, especially with staff. One extended session involved attending a recreation planning meeting with a resident and two staff. Two additional 20-minute interviews with residents were conducted by telephone. They were unstructured interviews, using open-ended questions that followed the leads provided by the first general opening, "Can you tell me about the courtyard space at Redwood Gardens?"

FINDINGS

The courtyard, then and now, is loved and appreciated. In 1994 residents said that they liked the space. However, the students observed that it was underused. At the time, we reasoned that this was because it is bisected by a pathway, did not have enough shade, did not have enough privacy, and needed more movable seating. The most actively used part of the courtyard was the allotment gardens, although approximately 25% looked fallow, with little or no activity in evidence. One resident from the 1994 study summarized the problems with different features of the courtyard:

> The tables in the amphitheater are nice, but I can't use it because there are no trees and it gets too sunny. Up above in the courtyard, where the trees are, there is too much traffic. You can't sit there and read a book or talk because there are always people walking by you. There is no privacy in the courtyard.

In 2005 residents use adjectives like "great," "beautiful," "quiet," "pleasant," "peaceful," and "meditative," to describe the courtyard. When we asked residents directly if the courtyard is used, many would say that it is used 'a lot.' But we ourselves did not see people gardening, nor people sitting to chat or read. "Come later in the afternoon," we were told; "come on a sunny day." Still, even at those times the courtyard could not be characterized as a beehive of activity (Figure 3). What accounts for this discrepancy between the sincere appreciation of the space and its low usage? In the balance of this paper we consider 10 different explanations for this discrepancy. To what extent is design responsible for both the appreciation and the under use? We conclude by discussing the design implications of each possible explanation.

The Population Is Too Active to Need the Courtyard for Exercise

Most of the residents are still active. Only 10% of the units were built following the guidelines for universal design, so the remaining 90% of the units require that the residents be able-bodied, one of the reasons being that the doorways are not wide enough for wheelchairs. Anyone who has an accident or becomes infirm enough to require a wheelchair has to move out of the building altogether. These active residents appreciate the courtyard but do not particularly need to use it because they are active in the neighborhood and beyond. Furthermore, there are only 188 residents in 168 units (HUD, 2005), so it is not realistic to expect a lot of people at any one time in the courtyard, especially considering that many are out in the daytime.

The Residents Do Not Feel a Need to Leave Their Private Units

According to residents interviewed, the private units are satisfying. When residents are at home they are happy to stay in their units. The courtyard is viewed as a bonus, not as a compensation for poor living conditions.

FIGURE 3. Even this major pedestrian thoroughfare is often empty.

The Courtyard Is Too Public for Some Activities and Moods

Those residents who feel under the weather do not like the idea of going out into the more public space of the courtyard. Yet these are the residents who may potentially benefit the most from an hour spent outdoors, according to recent research that shows that "having a one-hour rest outdoors in a garden setting plays a role in elderly people's powers of concentration, and could thereby affect their performance of activities of daily living" (Ottosson & Grahn, 2005). We are left with a question about how to get this ailing part of the population to use the courtyard. Creating more semi-private subdivisions with the feeling of visual privacy would be one solution.

Weather Influences Patterns of Use

The courtyard is only used in good weather; even then it is only used early or late in the day when the sun is not too direct and hot. Even in the

early spring when the temperature was only in the upper 60s, the courtyard at Redwood Gardens gets very hot when the sun is high, so we
were advised to come back at 5:30 PM. A flat cement pathway around
the perimeter of the western side of the courtyard functions as a kind of
track for those who want to walk. We were told that some people prefer
to walk after dinner when the sun is down. Similarly, some work in their
allotment gardens in the early evening. We were told that a warm Friday
evening generates the most activity, so we went to observe for ourselves
on a warm Friday evening. However, despite more residents being in
the courtyard, their numbers were modest, 5 residents and 2 of their regular visitors. Providing shade could increase use during the entire day
(Figure 4).

We learned that if the sun is too hot, or the weather too cold or rainy,
some of the residents use the corridors inside the buildings as a kind of
circuit for walking. The fact that people can walk indoors means that the
interior planning is especially adequate and has health benefits that
could be replicated elsewhere at other sites. When the weather is bad,

FIGURE 4. Many wait until the sun is down to use this walking track.

two organized exercise groups, tai chi and yoga, use the community room. This underscores the importance of indoor as well as outdoor space for physical exercise. Redwood Gardens offers a good precedent for new projects which could benefit from indoor circulation.

Direct Use Is not Important; Visual Use Is Enough

Several of our informants did not mind that the courtyard is not used a lot. They said that they "just like knowing it's there." It is viewed as an amenity, which raises this housing above the minimum to be "country club." Having given up their own yards and gardens, many still like having some garden to show visitors, being one of the main places to take visitors. One woman described taking her visiting friend there to sit under the shade of an umbrella to play Scrabble, others reported that visiting children would be taken there. Several individuals take pride in tending to some particular feature of the courtyard landscaping.

From the interviews with residents we have concluded that they benefit from the courtyard even if they do not use it physically. Everyone was aware that this was an extra feature that others in the same position (in publicly assisted housing) do not have. Housing specialists would consider the courtyard, an amenity, not an essential. But, from the point of view of mental health (self-esteem), it certainly is beneficial. This psychological boost is likely to improve physical health. Insofar as mental health is important, a resident does not have to use the courtyard physically to benefit from it.

Noisy Activities Are not Allowed in the Courtyard

Activities in the courtyard are thought to help create community, but some residents want only quiet activities there. Residents who live facing the courtyard are able to hear the echo of whatever is going on there. Resident experience suggests that noise is amplified in the courtyard. Acceptable activities include walking, reading, conducting reading groups, and allotment gardening. The courtyard is appreciated as a place to get exercise close to home without having to go away. Thus, walking on the walking tract is a particularly valuable activity, quiet enough not to disturb residents.

Noise is a big issue, as shown by seven residents having petitioned the management to prohibit the use of gas-powered mowers by landscape maintenance crews. They were successful because they proved that the noise level violated Berkeley's noise ordinance. We were told that a victory sign went up in the elevator saying, "We can now enjoy the silence of the lawns." On one day of the year–the Fourth of July–no one worries about noise. Several of our 2005 interviewees mentioned the use of the courtyard for the Fourth of July party when tables are set up for a barbeque.

Cultural Misunderstandings Might Dampen Usage

Noise is an important factor in deciding what activities occur in the courtyard, but not the sole factor. Some types of noise were more acceptable than others. The noise made by one group talking was considered acceptable, but another talkative group was considered noisy. In good weather a group does tai chi in the courtyard at 8 AM daily, but their conversation, usually in Chinese, has disturbed some residents. In contrast, an art class once used the courtyard, but the woman who runs it moved the class into her apartment for health reasons. When asked if the class made too much noise, one informant said, "I can hear them, but I liked to go out and see what they were doing. I also make art in the courtyard." In contrast, the same woman found the Chinese language ". . . a difficult language to listen to if you are not use to it." Therefore, the tai chi group was noisy to her, but the art group was not. This small example of cultural misunderstanding points to another potential problem in the use of common outdoor space. Intercultural differences could be a factor in limiting activity in common spaces.

At Redwood Gardens 58% of the residents are Caucasian, 24% are Chinese and other Asian ethnicities and Pacifica Islanders, 5% are Hispanic, 1% are American Indian/Alaskan Native, and 12% are African-American (HUD, 2005). By and large, this multicultural population co-exists peacefully. But 2 or 3 women of European descent lowered their voices when they acknowledged to us, one Caucasian and one Asian, that many Asians live in the complex. Successful public spaces allow different kinds of people to coexist even if they do not interact with one another.

Allotment Gardens Attract Attention

Tending the garden plots is a quiet activity (Figure 5). The allotment gardens were used to grow vegetables, flowers, and herbs. These small plots were used more than the other features of the courtyard, including seating along the major path, amphitheater, and lawn area. The waiting list for garden plots indicates their popularity. Residents have some other possibilities for gardening. One woman on the waiting list asked to use a corner located in the gardening area for tending potted plants, and was given permission. Another woman plants flowers in the tree planters and many use a side patio outside of the courtyard to keep potted plants (Figure 6). Another woman tends a "secret garden" in a hidden location.

The Courtyard Is too Public Without Enough Territories

The courtyard is too open and undifferentiated for psychological ease-of-use. The transition between public and private needs to be more

FIGURE 5. Allotment Gardens

gradual. One common professional design guideline is that the transition between housing in the public realm should have at least four phases: private, semiprivate, semi-public, and public (Marcus, 1986). Redwood Gardens offers these four kinds of spaces (units, corridors, courtyard, bus stop at the street), but more are needed. The courtyard itself needs to be further differentiated. The pathway crossing from one side of the courtyard to another is semi-public; what is missing is more semi-private niches or territories in the two halves of the courtyard. The most utilized area of the courtyard is a lounge area facing south at the north entry. It is protected by walls and railings, defined by an indentation in the building, and completed by a railing between it and the garden (Figure 7).

Part of walking is having an attractive destination, and here seating plays an important role. The central pathway would benefit from some refinement with regard to seating. See the discussion of policy and design recommendations below.

FIGURE 6. Side Patio with Potted Plants

In contrast to the successful allotment gardens, the least used portion of the courtyard is the "amphitheater," a cement hemisphere surrounded by sloping lawn (Figure 8).

The architects use the term amphitheater, but some residents refer to it as the "flat part." They claim that only visiting children who play ball on the cement surface use this area. In 1994 one resident complained, "The tables in the amphitheater are *nice*, but I can't *use* it . . ." This area would benefit from being redesigned. One informant referred to the area as an "acoustic shell," recognizing that its hard surface amplifies the sounds that disturb residents. The redesign of this central area, ideally around some particular activity or purpose, should include soft earth, vegetation, a sense of containment, and shade.

Residents take special delight in a "secret garden" between the windows and hedges (Figure 9). It is not literally secret, but it is screened from view. A woman took advantage of a 2-foot space between her windows and shrubbery that had been planted to provide that unit with visual privacy to create a narrow garden and aviary. It is an eloquent

FIGURE 7. Seating and Retaining Wall with Railing

FIGURE 8. Amphitheater or "Flat Area"

testimony to the need for small territories created through screening or territorial marking.

One resident in 1994 succinctly summarized the need for more privacy: "Up above the courtyard, where the trees are, there is too much traffic. You can't sit there with a book or talk because there are always people walking by you. There is no privacy in the courtyard." His statement supports Rapoport's situational definition of privacy, which is not an absolute physical condition, but rather a sense of control over unwanted interaction with others (Rapoport, 1975).

No Smoking Policy Might Direct Some Residents Elsewhere

Smoking within the courtyard has been prohibited, presumably because having buildings on all four sides means that wind cannot move and clean the air continuously. Those who might have used the space to smoke have to go elsewhere. This is a good policy from a health point of view for smokers and nonsmokers alike and should be continued.

FIGURE 9. The Secret Garden

POLICY AND DESIGN RECOMMENDATIONS

This assessment of the factors with the potential to affect courtyard usage reveals that design is not entirely responsible for its use or disuse. However, insofar as physical design *is* responsible, we can offer the following suggestions. Note that these recommendations are for Redwood Gardens to assist its residents and management, but they can usually be applied to housing for the elderly more generally in order to assist planners and designers elsewhere.

1. The fact that so many residents are able-bodied means that some residents are active enough to participate in refurbishing and enriching the landscaping of the courtyard (Figure 10).
2. Even those who leave the site to participate in community life elsewhere would appreciate improvements to landscaping because it is valued as special to this housing project.
3. Sub-territories are needed in almost all common open-spaces, this courtyard in particular, so that people feel comfortable going to "their spot" even if they are feeling vulnerable when not well. Ways to achieve such niches are discussed further in point 9 below.
4. The influence of weather means there is a need for shade. Some kinds of shades could also protect from rain. But such structures should be limited and small, and sensitively placed so as not to block the light or views from units on the first-floor. Movable shade umbrellas, deciduous trees, and vine-topped pergolas would be appropriate to the task. Providing circuits for walking indoors and planning community rooms large enough for exercise groups should be a part of planning elsewhere.
5. Even for those who only use it visually, landscaping is a source of pride. However, it is getting a bit spotty, so it would benefit from infill plantings (Figure 11a) (Figure 11b). If there is no budget for such improvements community gardeners could take on this responsibility, given that many are physically fit enough for gardening. Additionally, a resident group, the management, or both, could make a suggestion to residents to let their friends and family know that living shrubs and plants would be especially appreciated on special occasions, rather than the cut flowers that are customary.
   From a visual point of view, making the landscape more full and abundant would be satisfying to residents and visitors in its own right. Additionally, judicious use of plantings could help form the semi-private niches needed for seating.

6. Noise management would benefit from softening the acoustic surfaces of the building walls that enclose the courtyard. Again, additional planting is called for. We recommend planting additional trees at the walls between windows so as not to obscure views. The species should be selected with awareness of roots being near building foundations. An additional benefit is that reducing the amount of cement surface might lessen noise reverberation.

The amphitheatre could be redesigned with a soft floor and more vertical enclosure with shrubs, light poles and trees. The cost of this proposal means that it does not have the high priority of other less expensive recommendations.

7. Providing defined territories that facilitates co-existence will maximize the potential for cultural acceptance and interaction.

8. Because of the popularity of allotment gardening and the direct physical activity it encourages, we recommend creating another strip of allotment gardens, and support continuing the policy of offering alternative spaces for those on the waiting list. Allotment gardening has been popular historically whenever and wherever offered–in public parks and vacant lots all over the United States (Cranz, 1982). Other housing projects would benefit by including allotment gardens in their site plans.

9. More niches and better-placed seating would likely increase use of Redwood Gardens' inner courtyard. The site planning guidelines about the number of transition zones between public and private should be revised upward elsewhere as well. At Redwood Gardens the courtyard should be subdivided, while maintaining a sense of the whole. No high cement walls are called for, but rather subtle screening and enclosure through plantings. Walls would only work if they were low–at table height. At this height they could also serve as high benches for short term perching.

The walking path around the perimeter is important to residents who emphasized the importance of having a smooth and level pathway. The slightest irregularity in the surface might make someone trip and fall, which runs the risk of breaking bones. The elderly are acutely aware of the danger of breaking a hip, and associated complications which too often lead to death. The perimeter path is one of the ways in which a sense of the whole will be maintained in the courtyard (See Figure 4).

FIGURE 10. Resident Clipping Shrubs

FIGURE 11a. Spotty Landscaping

FIGURE 11b. Improved Landscaping

More eddies off the side of the main pedestrian thoroughfare that connects the two long buildings would help invite residents outdoors (See Figure 3). Such eddies allow the sitter to watch others without being in the center of the flow. Improving the placement of seating is relatively easy and inexpensive with a huge potential for increasing psychological ease of use. The bench on the promontory formed by the walkway is poorly placed because the sitter has her back to the flow of traffic and the view is of a relatively empty part of the courtyard. The Chinese would consider this bad *feng shui*. (Meaning literally "wind and water" *feng shui* refers to the art and science of placement–of cities, buildings, cemeteries, and objects.) The orientation of the bench should be reversed 180°. One might rightly argue that having one's back to a set of stairs is also unsettling psychologically. However, in this case the hand railing behind the bench would provide a sense of certainty that the bench cannot tip over backwards down the stairs (Figure 12). Another bench in one of the front courtyards is unused because it sits at the bottom of a slope! It should probably be removed since reorienting it would only put a person's back to the thoroughfare, the psychologically uncomfortable position just discussed (Figure 13).

The designers lost an opportunity to use a retaining wall for short-term informal sitting and perching (Figure 14). In future projects archi-

FIGURE 12. Sitting with your back to passersby is never comfortable, and the view in this case does not merit the strain; this can be fixed easily by turning it the other way.

tects and landscape architects elsewhere should not place the railing in a retaining wall so that no space is left for comfortable short-term sitting.

AMENITIES AND AFFORDABLE HOUSING

User-oriented evaluations are not the sole criteria for judging the success of every feature of the building. For example, the architect was particularly proud of being able to persuade the HUD bureaucracy to approve the use of a red tile roof, to fit in with the surrounding context, even though tile would ordinarily be prohibited as an unnecessarily expensive amenity. No residents spontaneously singled out this feature as

FIGURE 13. A bench outside of the central courtyard is uncomfortable physically and psychologically at the bottom of a slope, even though the orientation to the pathway is correct.

the reason for their overall housing satisfaction. Yet it undoubtedly plays an important, if subliminal, part in their pleasure in living in an environment that is so "nice," more like a "country club" than "just housing."

While we want to acknowledge that residents may not be able to assess everything explicitly, we also want to emphasize that we have learned a substantial amount by taking the time to observe and to ask residents about their experience of living here. Culturally, congregate housing for the elderly is still a relatively new building type and our understanding of what makes shared outdoor spaces work psychologically, socially, and culturally is still evolving. We have made a modest contribution to this understanding with what we have learned about the importance of shade, physical activity, and creating more territories within common spaces.

FIGURE 14. This retaining wall could have provided improvisational seating along its length.

REFERENCES

Brummett, W. J. (1994). *The Essence of Home: Architectural Design Consideration for Assisted Living Elderly Housing.* Milwaukee, WI: Center for Architecture and Urban Planning Research, University of Wisconsin.

Carstens, D. Y. (1985). *Site Planning and Design for the Elderly: Issues, Guidelines, and Alternatives.* New York, NY: Van Nostrand Reinhold Company.

Cranz, G. (1982). The Politics of Park Design: A History of Urban Parks in America. Cambridge, Mass.: MIT Press.

Cranz, G. (1994). *Redwood Gardens Post Occupancy Evaluation, Spring 1994, Architecture 110.* Berkeley, CA: University of California Berkeley.

Cranz, G., & Christensen, D. L., & Dyer, S. (1976). *A User-Oriented Evaluation of San Francisco's Public Housing for the Elderly.* Berkeley, CA: Department of Architecture, University of California.

Cranz, G., & Schumacher, T. L. (1975). *Open Space for Housing for the Elderly.* Princeton, NJ: Research Center for Urban and Environmental Planning, School of Architecture and Urban Planning, Princeton University.

Hoglund, J. D. (1985). *Housing for the Elderly: Privacy and Independence in Environments for the Aging.* New York, NY: Van Nostrand Reinhold Company.

Lawton, P. M. (1987). *Planning and Managing Housing for the Elderly.* New York: Wiley.

Marcus, C. C. (1986). *Housing as if People Mattered: Site Design Guidelines for Medium-Density Family Housing.* Berkeley, CA: University California Press.

MicroHUD/RENT 200 Redwood Gardens Coop. (2005). *Family and Head of Household Statistics.* (121EH153).

Ottoson, J., & Grahn, P. (2005). A Comparison of Leisure Time Spent in a Garden with Leisure Time Spent Indoors: On Measures of Restoration in Residents of Geriatric Care. *Landscape Research,* 30(1): 23-55.

Pastalan, L. A., & Schwarz, B. (2001). Housing Choices and Well-being of Older Adults. *Journal of Housing for the Elderly.* New York: Haworth Press.

Rapoport, A. (1975). Towards a Redefinition of Density. *Environment and Behavior,* June: 138-155.

Regnier, V. A. (1985). *Behavioral and Environmental Aspects of Outdoor Space Use in Housing for the Elderly.* Los Angeles, CA: Andrus Gerontology Center, University of Southern California.

Regnier, V. A. (1994). *Assisted Living Housing for the Elderly: Design Innovations from the United States and Europe.* New York: Van Nostrand Reinhold.

Regnier, V. A., and Hamilton, J., & Yatabe, S. (1995). *Assisted Living for the Aged and Frail: Innovations in Design, Management, and Financing.* New York: Columbia University Press.

Regnier, V. A., & Pynoos, J. (1987). *Housing the Aged: Design Directives and Policy Considerations.* New York: Elsevier Science Publishing Co.

Schwarz, B., & Brent, R. (1999). *Aging, Autonomy, and Architecture: Advances in Assisted Living.* Baltimore, MD.: John Hopkins University Press.

Stoneham, J., & Thoday, P. (1994). *Landscape Design for Elderly & Disabled People.* Chichester, U.K.: Packard Publishing.

U.S. Department of Health and Human Services. (2001). *The Surgeon General's Call to Action to Prevent and Decrease Overweight and Obesity.* Washington, DC.: U.S. GPO.

Resident Perceptions of Physical Environment Features that Influence Outdoor Usage at Assisted Living Facilities

Susan Rodiek

SUMMARY. Spending time outdoors has been found to have therapeutic potential, and to be highly valued by older adults, yet outdoor areas at residential care facilities are commonly reported as being underutilized. To learn what features of the physical environment were perceived by residents as either attracting or deterring outdoor usage, this study conducted focus groups and written surveys at fourteen assisted living facilities, with 108 residents ranging from 61 to 97 years of age. Participants reported that their outdoor usage was influenced by accessibility, aesthetic concerns, and specific environmental features such as the provision of shade, seating elements, plants, and views. The rural vs. urban

Susan Rodiek, PhD, NCARB, is Associate Director of the Center for Health Systems & Design, and Assistant Professor in the Department of Architecture, both at Texas A&M University, College Station, TX.

Address correspondence to: Susan Rodiek, PhD, Department of Architecture, MS 3137, Texas A&M University, College Station, TX 77843-3137 (E-mail: rodiek@tamu.edu).

Research for this article was supported in part by a grant from the College Research and Interdisciplinary Council, College of Architecture, Texas A&M University.

[Haworth co-indexing entry note]: "Resident Perceptions of Physical Environment Features that Influence Outdoor Usage at Assisted Living Facilities." Rodiek, Susan. Co-published simultaneously in *Journal of Housing for the Elderly* (The Haworth Press, Inc.) Vol. 19, No. 3/4, 2005, pp. 95-107; and: *The Role of the Outdoors in Residential Environments for Aging* (ed: Susan Rodiek, and Benyamin Schwarz) The Haworth Press, Inc., 2005, pp. 95-107. Single or multiple copies of this article are available for a fee from The Haworth Document Delivery Service [1 800 HAWORTH, 9:00 a.m. - 5.00 p.m. (EST). E-mail address: docdelivery@haworthpress.com].

Available online at http://www.haworthpress.com/web/JHE
© 2005 by The Haworth Press, Inc. All rights reserved.
doi:10.1300/J081v19n03_06

background of participants was compared with self-reported time spent outdoors, and no significant association was found. *[Article copies available for a fee from The Haworth Document Delivery Service: 1-800-HAWORTH. E-mail address: <docdelivery@haworthpress.com> Website: <http://www. HaworthPress.com> © 2005 by The Haworth Press, Inc. All rights reserved.]*

KEYWORDS. Aging, elderly, outdoor environment, nature, long term care, residential care, underusage, older adults

INTRODUCTION

Research increasingly suggests that spending time outdoors may have therapeutic benefits for older adults, through pathways such as increased physical activity, appreciation of nature elements, enhanced sense of autonomy, and the hormonal effects of bright outdoor light levels (Kono et al., 2004; Namazi & Johnson, 1992; Takano et al., 2002; Ulrich; 1999). Although exterior space may be perceived as a therapeutic resource for residents of long term care facilities, existing outdoor areas are widely reported to be underutilized (e.g., Heath & Gifford, 2001; Hiatt, 1980). In practice settings, underusage may be attributed to factors such as resident disinterest, poor health, unfavorable climate, staff protectiveness, or urban background of residents (e.g., Humpel, Owen, & Leslie, 2002). Although the impact of these factors on outdoor usage has not been thoroughly explored, it seems likely that they would influence usage in many situations. Elements and configurations of the physical environment are considered to be another important influence on resident use of the outdoors (Taylor et al., 1993); these are amenable to environmental design intervention, and are therefore the focus of this study. A main purpose was to explore resident perceptions of the physical environment, in order to determine which elements were perceived as being either 'magnets' (attractants) or 'barriers' (deterrents) influencing their usage of outdoor spaces.

There appears to be little empirical evidence that comprehensively describes the relationship between physical environment features and the use of outdoor space by residents at long term care facilities. A few findings on this topic have emerged as 'additional information' from studies with a non-related emphasis. For example, a recent national survey that examined reasons for and patterns of leaving assisted living facilities, in relation to the level of services offered (Hawes, 2002), found in their field surveys that residents tended to be less satisfied with outdoor spaces than with other parts of the facility environment. Both Hiatt (1980) and Heath and Gifford (2001) conclude that usage of outdoor spaces is strongly related to the design and planning of the physical environment. Design practitioners have also criticized aspects of facility design as an important reason for underutilization of outdoor space (e.g., Carstens, 1982; Cranz, 1987; Robson et al., 1997). Perceived underusage contributes to the widespread belief that aging residents have little interest in the outdoors. By contrast, the literature suggests that elderly residents tend to regard nature contact as enjoyable, and as being valuable to their health and well-being, even if they spend minimal time outdoors at their facility (e.g., Cohen-Mansfield & Werner; 1998, MacRae, 1997; Perkins, 1998).

Two main goals of this study were: (1) to assess whether residents perceived environmental features as influencing their outdoor usage, and (2) to identify specific features they perceived as either attractants or deterrents to outdoor usage. Knowing this would make it possible to develop evidence-based guidelines that summarize the primary ways the environment can support outdoor usage by residents. The main hypotheses were that (1) the design and condition of the physical environment plays an important role in outdoor usage, and that (2) specific environmental features may act as magnets or barriers, that either encourage or discourage usage.

METHODS

Facilities and Participants

A multi-stage cluster sampling strategy was used, in which facilities were randomly selected, and then residents were randomly selected from each facility. Fourteen facilities were chosen in a twelve-county region of southeast Texas, from the list of assisted living facilities in that region registered by the state as having a capacity of fifty or more persons. Eleven of the fourteen facilities were located within the urban area or outer fringe of Houston, which is the only large city in this region; the remaining three facilities were located in medium-sized towns. Licensed capacity of facilities ranged from 57 to 188 residents; there was a fairly even mix of one-, two- and three-story buildings. From each facility, approximately eight residents were selected from lists of those considered by facility administrators as being physically and cognitively able to participate in the study. In total, 108 residents participated, with a mean age of 83.6, and ranging in age from 61 to 97 years, with approximately 65% being female. Approximately 10% reported their health as excellent, 51% good, 33% fair, and 6% poor.

Instruments and Pre-Testing

The study used a combination of written questionnaires and focus group discussions. It was deemed sufficiently redundant to conduct focus groups at seven facilities, while written surveys were conducted at all fourteen facilities. At sites where both methods were used, the written surveys preceded the focus group sessions, so individual responses would not be influenced by the group discussion. Both the survey questions and the focus group discussions addressed the same range of topics, with focus groups adding an interactive dimension. The study investigated how residents felt about going outdoors, what they did there, what environmental features they found appealing, and what features presented barriers to going outdoors.

Surveys took approximately ten minutes for residents to complete, and combined five open-ended and ten closed-ended questions. Pretesting of response formats indicated that higher numbers of response categories tended to cause confusion and lead to diminished validity of responses. With older adults, Likert-type scales have been found to be less stable than simple binary (yes-no) questions (Kane et al., 2003), possibly because scales require greater mental effort. Lack of response variability has also been found to be a problem with older adults, who tend to use mainly the upper end of the response format, even with a large number of response categories (Castle & Engberg, 2004). In this study, simplified versions of Likert-type scales were used, with descriptive words to articulate each choice. Although seven-point Likert scales are often used with younger subjects, this survey used a four-point scale. An even number of response categories was used, to clarify the findings by requiring residents to choose a direction of preference (Converse & Presser, 1986; Fink, 1995). Two of the items that otherwise might have led to frustration, provided a neutral response option. Several questions provided write-in space for additional comments. Questions were pre-tested and revised to reduce ambiguity, and were formatted with layouts designed to simplify the response process. The final version of the questionnaire consisted of three pages with oversized Arial type, with a cover sheet for identifying information.

Focus group sessions were conducted in groups of approximately seven to ten residents, with a moderator guiding the discussion. Sessions lasted about forty-five minutes to an hour, and followed a semi-structured progression of topics, from general to specific. Nearly 90% of participants were recorded as having made at least one contribution to the discussion. After three or four focus group sessions, recognizable themes began to emerge from the issues that were brought up repeatedly by residents. By the final session, most of what residents brought up had already been discussed in earlier sessions, and it was estimated that satisfactory saturation of the most important aspects of the topic had been reached, based on criteria described by Fern (2001) as a point of 'diminishing returns' (p. 162). Tape recordings and written notes were transcribed for later analysis. To establish an environmental context, the site plans of the seven facilities used in focus group sessions were measured, drawn to scale, analyzed, and annotated as a basis for understanding the results found in this study.

RESULTS

Analysis

Although the survey questionnaire was very brief, several respondents skipped some of the questions, and relatively few participants completed all the write-in responses. All readable responses were retained and the closed-ended survey responses were summarized. The open-ended survey responses were typically brief and could be unambiguously grouped into categories; these were then combined with the results of the focus group content analysis. These combined results are presented here as 'frequency of comments,' to indicate the number of times each topic was brought up by residents. This gives an approximate indication of how salient these topics were to participants during this process, as they responded to open-ended questions about their outdoor usage. Qualitative results were categorized according to criteria outlined by Jones (1985, p. 125; after Holsti, 1969), who recommend that categories for content analysis should:

1. reflect the purpose of the research,
2. be exhaustive,
3. be mutually exclusive,
4. be assigned independently, and
5. be derived from a single classification principle.

Categories developed with these criteria were found to cover nearly 95% of the comments, with the remainder being non-relevant topics. From these open-ended responses and focus group findings, the specific environmental features described by residents were organized into categories of similar elements.

Role of Physical Environment in Outdoor Usage

To elicit whether the physical environment was perceived as an overall barrier or support to outdoor usage, residents were asked, "Do the building or outdoor areas make it hard or easy for you to be outdoors?" This wording suggested that residents consider both indoor and outdoor elements as related to their outdoor usage. A write-in space below the question allowed them to describe the elements they associated with this ease or difficulty; these responses were sorted into the data for specific environmental features. Table 1 shows that 76% responded that the building *did not* make it hard to go outdoors, while 24% felt the environment *did* make it harder to go outdoors. The response rate was uniformly lower on all questions asking residents to speculate in unfamiliar ways (78% on this item), compared with approximately 95% response rate on easier-to-answer questions.

Barriers to Outdoor Usage: Issues Other than Accessibility

In spite of the fact that only about one-quarter of participants indicated that the physical environment made it hard for them to use the outdoors, many residents volunteered unfavorable comments on built

TABLE 1. Reported Ease of Outdoor Usage (n = 83)

Ease of Outdoor Usage	Frequency	Percent
"Do the building or outdoor areas make it hard or easy for you to be outdoors?"		
Very hard	2	2%
Somewhat hard	18	22%
Somewhat easy	37	45%
Very easy	26	31%
Total	83	100%

and natural elements (see Tables 2 and 3). Overall, the environmental conditions that were considered barriers to outdoor access were described both as the *presence of undesirable* elements, and the *absence of desirable* elements. Their concerns primarily related to how the built environment was planned, designed and/or maintained. Many of their comments had comfort components; for example, residents indicated they would like more adequate and closely spaced outdoor seating, and more protection from the sun and rain. There were numerous safety and security concerns, primarily regarding the safety of walking surfaces. In regard to the convenience of reaching outdoor areas, one resident commented, "Outside doors are not close to my room." Residents also indicated there was a lack of amenities such as swings, and interesting landscapes to look at. For example, one resident asked for "a place to walk where the scenery is different–a circle is not enough–it's monotonous." Another said he would enjoy using a place that was "like a small park," and had less concrete paving.

Barriers to Outdoor Usage: Accessibility-Related

It was surprising that many of the unfavorable comments on built elements referred to accessibility issues, even though these were all modern facilities that generally appeared to have been built with accessibility in mind. Table 3 shows the main categories of accessibility-related problems that emerged from this study. Several problems appeared to develop from poor coordination between interior and exterior spaces, the deterioration of materials, and lack of adequate maintenance. For example, one paved sidewalk was sufficiently wide in itself, but was intruded upon by the overhanging front ends of adjacent parked cars, so wheelchairs could not pass easily. At another fairly new facility, sec-

TABLE 2. Response Categories Reporting Problems with Outdoor Usage

Problems Reported With Outdoor Usage	% of responses
Physical configuration of elements	32%
Safety/security concerns	29%
Insects and/or climate conditions	23%
Lack of interesting features	16%
Total	100%

TABLE 3. Response Categories Reporting Outdoor Accessibility Issues

Accessibility Issues Reported with Outdoor Usage	% of responses
Problems w/sidewalks	42%
Problems w/doors	25%
Wheelchair usage	21%
Distance (too far)	11%
Total	100%

tions of an accessible sidewalk had settled and cracked after construction, thereby making it difficult to navigate with a walker or wheelchair. Doors were reported to be especially difficult to negotiate, and several residents reported that they seldom went out because "the doors are hard to open."

Magnets Attracting Outdoor Usage: Built Environmental Elements

Residents described both built and natural features as attracting them to use outdoor spaces, as shown in Tables 4 and 5. Many of the comments by residents suggested that different features of the *built* environment influenced their usage of outdoor areas, as shown in Table 4. Although some of the comments about built elements were favorable, the majority included some level of criticism, and many responses combined complaints and compliments in the same statement. Participants reported the need for more satisfactory overhead shelter, better walking paths, and comfortable places to sit outside. The two main categories of criticisms of built elements can be described as *lack of comfort*, and *lack of accessibility*. Because of the open-ended methodology used, the summary provided here is only an indication of the relative importance of these elements to residents, and does not clearly distinguish elements that were found to be satisfactory at their facility, from those that were unsatisfactory. This information could provide the basis for a further quantifiable assessment measure.

Magnets Attracting Outdoor Usage: Natural Elements

Appreciation of specific natural environmental features shown in Table 5 supports earlier findings (Perkins, 1998; Talbot & Kaplan, 1991), with fresh air, trees/green plants, and flowers being listed as the features residents liked best about the outdoors. In this study, several residents described what they most wanted as being like 'a small park' or a 'strolling garden,' where they could walk under trees and among attractive plantings without going far from the facility. Overall, natural features were typically described as 'magnets' that increased the interest of residents in going outdoors, shown by comments such as "I miss trees," or "I love watching the birds." Elements of nature were clearly described as providing some of the main reasons residents went outdoors, and were the elements appreciated most while using outdoor areas. Occasionally, natural features were described as being uncomfortable, such as (the) "sun is too hot on my skin."

TABLE 4. Response Categories Reporting Preferred Built Elements

Preferred Features of the Built Environment (*primarily hardscape elements*)	% of responses
Overhead shelter	26%
Sitting areas	26%
Porches	18%
Gazebos	12%
Walking loop	8%
Swings	6%
Indoor features	4%
Total	100%

TABLE 5. Response Categories of Preferred Natural Elements (Includes Constructed Landscape Elements)

Preferred Natural Features	% of responses
Greenery	30%
Fresh air	18%
Flowers	17%
Birds	13%
Water features	9%
Other nature elements	6%
Sunshine	4%
Animals	3%
Total	100%

Urban-Rural Background Associated with Outdoor Usage

Because of the apparently widespread perception that people from urban backgrounds tend to be less interested in nature and the outdoors, this study also collected information on differences in urban-rural background of residents. Comments suggesting this possibility have been heard from residents, facility administrators, and other researchers–seemingly based on an assumption that rural background predisposes to an affinity for being outdoors. For example, one resident said, *"Of course* I like to be outdoors–I was raised on a farm. I'm a country girl." Although this seems plausible, in this study the self-reported time spent outdoors was found to correlate slightly negatively with rurality of background (see Table 6), although the difference was not significant (the large standard deviations in Table 6 reflect the large variation in outdoor usage found between residents).

DISCUSSION

This study emphasized the importance of safety and comfort issues, and confirmed preference for specific outdoor features that had previously been found to be meaningful to older adults. Study participants suggested that greenery, flowers, wildlife, and water elements were all attractants to outdoor usage. They also indicated that 'constructed' comfort features such as paved walkways and shade structures made it easier to spend time outdoors. Although residents expressed strong opinions about specific elements, the percentage who reported that the physical environment influenced their outdoor usage was lower than expected, and may reflect: (1) retribution bias, and/or (2) unfamiliarity

TABLE 6. Urban/Rural Background Correlated with Time Spent Outdoors (n = 105)

Frequency and Percent of residents in category		Childhood Background ("where I grew up")	Outdoor hours per week*		
			Mean	Median	SD
	31	29%	Large City	2.6	1.4	3.1
	49	47%	Medium Town	1.8	.3	2.6
	25	24%	Rural/Agricultural	1.5	.8	1.9
N	105	100%	Overall	1.9	.8	2.6

*calculated from reported frequency of usage and amount of time spent outdoors

with environmental design. This item had been anticipated to be difficult to assess directly, for two reasons: First, institutionalized older adults may be reluctant to criticize their facility, due to a tendency toward compliance and/or the fear of retribution (Smith, 2000). Second, it seems plausible that people unfamiliar with environmental design may consider the built environment to be expensive or difficult to change, and therefore tend to accept rather than criticize environmental conditions. These issues were addressed to some extent by telling residents in advance that the study's purpose was not to identify faults in their own facility, but to give their ideas on the best possible way to design future facilities.

An unexpected outcome of this study was that it found no significant difference in outdoor usage between residents who grew up in rural or in urban backgrounds. As this finding apparently contradicts a commonly-held assumption, further studies using methods such as in-depth interviews should be used to probe the issues involved, before drawing conclusions based on these preliminary findings.

Another issue that was not found to be as important as expected, was that residents might be concerned about maintaining visual contact with facility staff. Before the study, residents were hypothesized to feel some level of anxiety when venturing beyond the visual surveillance of staff; however, this did not emerge from the recorded responses. Comments by several residents suggested that after they reached the outdoors, most felt fairly comfortable being 'on their own' for a little while. Perhaps going out independently tended to foster feelings of autonomy, or perhaps the more autonomous residents tended to go out more. Because of the qualitative categorization of the results, it cannot be determined whether these comments reflect the feelings of the majority of residents in the sampled population. As this issue has been proposed as potentially important (e.g., Carstens, 1993), it should be further explored with more quantifiable measures.

This exploratory study investigated how residents perceived physical environment elements in relation to their usage of outdoor spaces. Based on these and other relevant findings, it would be worthwhile conducting in-depth interviews and additional closed-ended surveys to more definitely quantify the salient issues influencing outdoor usage, from the residents' perspectives. Further data on this important question could substantially strengthen the existing evidence base, the findings could help facilities make decisions on allocating scarce resources and prioritizing the improvement of outdoor spaces over a period of time. If research-based information were effectively translated into practice settings, the outdoors could potentially become a valuable therapeutic resource to improve residents' health and well-being, instead of remaining inadequately utilized.

REFERENCES

Carstens, D. Y. (1982) Behavioral research applied to the redesign of exterior spaces: Housing for the elderly, *EDRA: Environmental Design Research Association* (13), 354-369.

Carstens, D. Y. (1993) *Site Planning and Design for the Elderly: Issues, Guidelines, and Alternatives*. John Wiley & Sons, New York.

Castle, N. G. and Engberg, J. (2004) Response formats and satisfaction surveys for elders, *The Gerontologist*, (44), 358-367.

Cohen-Mansfield, J. and Werner, P. (1998) Visits to an outdoor garden: Impact on behavior and mood of nursing home residents who pace. In *Research and Practice in Alzheimer's Disease Intervention in Gerontology* (eds. Vellas, B. J., Fitten, G. and Frisconi, G.) Serdi Publishing, Paris, pp. 419-436.

Converse, J. M. and Presser, S. (1986) *Survey Questions: Handcrafting the Standardized Questionnaire*, Sage Publications, Beverly Hills, CA.

Cranz, G. (1987) Evaluating the physical environment: Conclusions from eight housing projects. In *Housing the Aged: Design Directives and Policy Considerations* (eds. Regnier, V. and Pynoos, J.) Elsevier Science, New York, pp. 81-104.

Fern, E. F. (2001) *Advanced Focus Group Research*, Sage Publications, Thousand Oaks, CA.

Fink, A. (1995) *How to Ask Survey Questions*, Sage Publications, Thousand Oaks, CA.

Hawes, C. (2002), Personal communication containing summarized field notes on national study (unpublished), December 6, 2002.

Heath, Y. and Gifford, R. (2001) Post-occupancy evaluation of therapeutic gardens in a multi-level care facility for the aged, *Activities, Adaptation, & Aging*, (25), 21-43.

Hiatt, L. G. (1980) Moving outside and making it a meaningful experience. In *Nursing Homes: A Country and Metropolitan Area Data Book*. U.S. Government Printing Office: Public Health Services Publication #2043. National Center for Health Statistics, Rockville, MD.

Holsti, O. R. (1969) *Content Analysis for the Social Sciences and Humanities*, Addison-Wesley, Reading, MA.

Humpel, N., Owen, N. and Leslie, E. (2002) Environmental factors associated with adults' participation in physical activity, *American Journal of Preventive Medicine*, (22), 188-199.

Jones, R. A. (1985) *Research Methods in the Social and Behavioral Sciences*, Sinauer Associates, Sunderland, MA.

Kane, R. A., Kling, K. C., Bershadsky, B., Kane, R. L., Giles, K., Degenholtz, H. B., Liu, J. and Cutler, L. J. (2003) Quality of life measures for nursing home residents, *Journal of Gerontology: Medical Sciences*, (58A), 240-248.

Kono, A., Kai, I., Sakato, C., and Rubenstein, L. Z. (2004) Frequency of going outdoors: A predictor of functional and psychosocial change among ambulatory frail elders living at home, *Journal of Gerontology: Medical Sciences*, (59A), 275-280.

MacRae, S. (1997) *Consumer perceptions of the healthcare environment: An investigation to determine what matters (Section: Long term care)*, The Center for Health Design and the Picker Institute, Martinez, CA.

Namazi, K. H. and Johnson, B. D. (1992) Pertinent autonomy for residents with dementias: Modification of the physical environment to enhance independence, *American Journal of Alzheimer's Disease and Related Disorders & Research*, (7), 16-21.

Perkins, S. D. (1998) The value of nature and the outdoors for older adults in congregate living facilities, unpublished Master's thesis, Architecture, Texas A&M University.

Robson, D., Nicholson, A., and Barker, N. (1997) *Homes for the Third Age: A Design Guide for Extra Care Sheltered Housing*, E&FN Spon, London.

Smith, M. A. (2000) Satisfaction. In *Assessing Older Persons: Measures, Meaning, and Practical Applications* (eds. Kane, R. L. and Kane, R. A.) Oxford University Press, New York, NY, pp. 261-299.

Takano, T., Nakamura, K. and Watanabe, M. (2002) Urban residential environments and senior citizens' longevity in megacity areas: The importance of walkable green spaces, *Journal of Epidemiology and Community Health*, (56), 913-8.

Talbot, J. F. and Kaplan, R. (1991) The benefits of nearby nature for elderly apartment residents, *International Journal of Aging and Human Development*, (33), 119-130.

Taylor, L., Whittington, F. J., Hollingsworth, C., Ball, M., King, S., Patterson, V., Diwan, S., Rosenbloom, C., and Neel, A. (2003) Assessing the effectiveness of a walking program on physical function of residents living in an assisted living facility, *Journal of Community Health Nursing*, (20), 15-26.

Ulrich, R. S. (1999) Effects of gardens on health outcomes: Theory and research. In *Healing Gardens: Therapeutic Benefits and Design Recommendations* (eds. Cooper Marcus, C. and Barnes, M.) John Wiley, New York, pp. 27-86.

Accommodating Culturally Meaningful Activities in Outdoor Settings for Older Adults

Susana Martins Alves
Gowri Betrabet Gulwadi
Uriel Cohen

SUMMARY. A growing multicultural aging population necessitates an examination of the cultural responsiveness of American healthcare environments in enabling successful aging experiences. Environment-behavior studies establish positive effects of natural environments on the

Susana Martins Alves, PhD in Architecture, is affiliated with Universidade Federal do Rio Grande do Norte, Departamento de Psicologia, Campus Universitário-Lagoa Nova, Natal, RN, Brazil 59078-970 (E-mail: susanamalves@terra.com.br). Gowri Betrabet Gulwadi, PhD in Architecture, is affiliated with Department of Design, Textiles, Gerontology & Family Studies, University of Northern Iowa, Cedar Falls, IA 50614-0332 (E-mail: betrabet@uni.edu). Uriel Cohen, D. Arch., is affiliated with School of Architecture and Urban Planning, University of Wisconsin-Milwaukee, Milwaukee, WI.

The authors would like to thank Marleen Carton from Eindhoven University, a visiting fellow at the Institute on Aging and Environment, School of Architecture and Urban Planning, University of Wisconsin, Milwaukee, for her work on the review of literature. The authors also wish to thank the participants in the study, and the facilities in which the interviews were conducted.

This study was funded by the USDA Forest Service through Grant Number 00-DG-11244225-205 and sponsored by the National Urban and Community Forestry Advisory Council.

[Haworth co-indexing entry note]: "Accommodating Culturally Meaningful Activities in Outdoor Settings for Older Adults." Alves, Susana Martins, Gowri Betrabet Gulwadi, and Uriel Cohen. Co-published simultaneously in *Journal of Housing for the Elderly* (The Haworth Press, Inc.) Vol. 19, No. 3/4, 2005, pp. 109-140; and: *The Role of the Outdoors in Residential Environments for Aging* (ed: Susan Rodiek, and Benyamin Schwarz) The Haworth Press, Inc., 2005, pp. 109-140. Single or multiple copies of this article are available for a fee from The Haworth Document Delivery Service [1-800-HAWORTH, 9:00 a.m. - 5:00 p.m. (EST). E-mail address: docdelivery@haworthpress.com].

Available online at http://www.haworthpress.com/web/JHE
© 2005 by The Haworth Press, Inc. All rights reserved.
doi:10.1300/J081v19n03_07

well-being of older adults. This study explored cultural differences in nature-related activities among Hispanic and Anglo-American elderly using six photographs of nature settings to elicit nature-related preferences and experiences. Findings indicate that Hispanic elderly find "furnished" natural settings more compatible with their preferred activities such as group-oriented socializing while Anglo-American elderly prefer "authentic" natural settings for preferred activities such as quiet reflection. Culture specific programming and design implications are discussed. *[Article copies available for a fee from The Haworth Document Delivery Service: 1-800-HAWORTH. E-mail address: <docdelivery@haworthpress.com> Website: <http://www.HaworthPress.com> © 2005 by The Haworth Press, Inc. All rights reserved.]*

KEYWORDS. Older adults, nature-related activities, culture, health care settings

INTRODUCTION

Aging and increasing ethnic and racial diversification are two main changes characterizing the population growth in the United States. Recent census data show that people of Hispanic or Latino (of any race) heritage make up 12.5% of the U.S. population. Whereas 15.0% of the Anglo-American non-Hispanic population is 65 years of age and over, approximately 5.0% of the Hispanic population in the United States is 65 years of age and over. There are nearly 1.7 million individuals living in 18,000 nursing homes in the United States–of the 1.733.591 million Hispanics (65 years and older), 37,680 individuals live in nursing homes whereas of the 29,244,860 White individuals (65 years and older), 1,388,783 live in nursing homes (U.S. Census Bureau, 2000). Growing numbers of an increasingly multicultural aging population requires the planning of suitable health care settings and programs that accommodate the residents' unique cultural assets or lifestyles.

An older person's ability to function adequately in health care environments depends on the disparities between the demands of the environment and the individual capacity to meet these demands. An individualized care approach aims to overcome some of the limitations of the medical model of care by focusing on the unique needs of individuals, their life history, values and traditions and the ability of the environment to provide for these needs. Within an individualized care approach, the environment must be designed and planned in a way that affords residents' participation and autonomy in decision making as well as opportunities to interact with others in meaningful ways.

An important aspect of relating the individualized care approach to design interventions is the consideration of a resident's unique cultural background. Sotomayor and Curiel (1988) emphasize that the Hispanic elderly group constitutes a heterogeneous population formed by a number of subgroups (e.g., Mexican-American, Cuban, Puerto Rican, Central or South American) that share a common cultural background not quite understood by policy makers and care providers in the U.S. In addition to the physical, psychological, and social changes that accompany the aging process, Hispanic people have to cope with the additional challenges of adapting to another culture (Sotomayor & Curiel, 1988). According to this "double-jeopardy" hypothesis, although Hispanic elderly are frequently more exposed to stress than the dominant culture of the host country, they use fewer coping resources due to their socioeconomic status, language barrier, and systematic exclusion from access to social and economic opportunities (Kessler & Cleary, 1980; Miranda, 1991).

Because cultural values influence the perception of access to, eventual use and consequent experiences in long-term care settings, we propose that it is important to examine pertinent cultural similarities and differences. Environmental interventions are less likely to be effective if they do not incorporate the issues most significant to the quality of life of the elderly; they should not only encompass general issues of cultural identity but also focus on activities specific to cultural groups and their concomitant social heritage. Therefore, in this study we focused on two groups–Spanish-speaking immigrant elderly and English-speaking non-Hispanic Anglo-American elderly.

Based on successful interventions in health care settings such as "healing gardens" and interaction with nature elements (such as Eden Alternative approach) that adopt an individualized care approach to maximize autonomy, choice, and continuation of lifestyles (Ulrich, 1999), we focused on the role of nature settings as suitable interventions. We examine how elderly users in two different ethnic groups perceive nature-related activities in their settings.

The goals of this study were:

1. to examine cultural differences in nature-related activities among elderly Spanish-speaking immigrants and English-speaking non-Hispanic Anglo-Americans;

2. to analyze the content of nature-related activities based on se-
    lected attributes of quality of life, such as engagement, a sense of
    control and social interaction;
3. to provide culturally sensitive recommendations for future na-
    ture-related health interventions in institutional settings.

A CONCEPTUAL FRAMEWORK TO GUIDE THE STUDY

This study links several dimensions together: the natural environment, quality of life of elderly residents in long term care environments,
and the role of culture. Cultural heritage is defined here as life style, activities, rituals, traditions, values, world view, language, material culture and other assets and habits shared by a group (Day & Cohen, 2000).
The role of cultural factors and ethnicity in person-nature relationships
is conceptualized as the filter, which assigns particular meanings to the
experience of nature, by a particular individual or a group of individuals
with shared cultural values. As illustrated in Figure 1, the experience of
nature of an individual within a cultural group is assessed using "cultural filters," and the resulting reactions–behavior, perception and
preferences–are influenced by cultural factors.

Another major premise of the study was that culturally-responsive
natural environment will lead, ultimately, to improved quality of life.
This improvement will be facilitated by more meaningful engagement,
opportunities for personal control and social interaction. The premise
above is derived from Lawton (1991) description of the four evaluative
sectors of quality of life: behavioral competence, perceived quality of

FIGURE 1. A Conceptual Framework of the Nature-Person Relationships from
a Multicultural Perspective (After Rapoport, 1977; Day and Cohen, 2000)

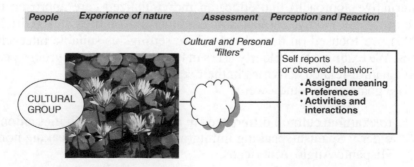

life, objective environment, and psychological well-being. The indicators of psychological well-being examined in this study are *engagement, a sense of control*, and *social interaction*.

Age-related biological and social changes alter the dynamics of one's engagement in activities, control over the environment, and one's level of social interaction. Despite these changes, the continuity of the self continues to be an operating human need. Our working assumption is that environmental interventions which facilitate the three components of psychological well-being–engagement, a sense of control and social interaction–would foster the continuation of the self. As helplessness, boredom, and loneliness are symptoms of decreased quality of life, we hypothesize that reductions in these problems through the use of culturally-responsive nature-related activities will increase elders' psychological well-being and foster the continuation of the self.

Figure 2 describes the application of the conceptual framework to our research design: Two cultural groups–Anglo-American and Hispanic–are the focus of the study. Culture is defined here as membership in an ethnic group; ethnicity is assumed to influence the nature of elderly people's activities. Day and Cohen (2000) provide a conceptual framework to address culturally relevant factors and their therapeutic value for the elderly, particularly those with Alzheimer's disease. They identify five main domains related to cultural heritage: history and life experiences, assets, beliefs and values, care giving practices, and activities and pref-

FIGURE 2. The Conceptual Framework of this Study: Selected Nature-Person Relationships from a Multicultural Perspective

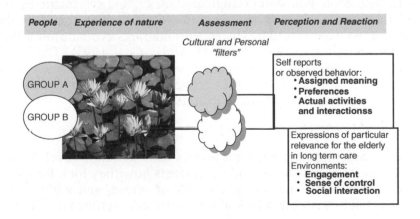

erences. The focus of the study is on the latter category–activities and preferences, as they relate to residents' engagement, sense of control and social interaction.

LITERATURE REVIEW

This literature review examines articles in two main domains:

1. Cultural Factors that Affect the Quality of Life of Elderly, and
2. Natural Environments as Settings Enabling Quality of Life Attributes–Engagement, Sense of Control, and Social Interaction

In the first part of the literature review, we focus on how cultural characteristics play an important role in determining the effectiveness of environmental interventions and caregiving approaches. Based on our conceptual framework and our assumption that aging creates a changing dynamic in the interaction between elderly and their environments, we focus on three particular symptoms of decreased quality of life in health-care institutions for the elderly; boredom, helplessness, and loneliness. In the latter part of the literature review, we examine nature settings as potential environmental interventions that enable these three quality of life attributes.

Cultural Factors Affecting the Quality of Life of Elderly

Elderly people's behavior and interaction with others is often marked by major cultural life experiences of the group. Fewer coping resources due to socioeconomic status, language barrier, and systematic exclusion from access to social and economic opportunities aggravate the level of stress experienced by Hispanic immigrants as ethnic minorities (Kessler and Cleary, 1980; Miranda, 1991; Cox, 1993). The stress caused by immigration has effects on how much control they will have over their environments, how they relate to others, and how they use public institutions and participate in activities.

Access and Use of Health Services by an Immigrant Culture

Anglo-American and Hispanic elders have different levels of access to health care services and this affects how they look for assistance in their social networks, country of origin, and within their families. Factors such as a lack of long-term care settings in minority communities, lack of health insurance, cultural differences between the providers of care and minority group members, cultural norms among Hispanics emphasizing children's responsibility for the care of aging parents, and cultural differences in living arrangements (Angel, Angel, McClellan, and Markides, 1996; Eribes and Bradley-Rawls, 1978) result in underutilization of long-term care services by Hispanic elderly (Angel, 1991).

Degree of acculturation–which relates to time spent in the mainstream culture since immigration, the group's country of origin, and the provision of community centers that support the groups' identity–has a significant impact on how elders seek care services. When living in Hispanic neighborhoods, Hispanic elderly tend to live close to their extended family to facilitate assistance and caregiving by members of the family. Assimilation can cause a greater distance between elders and their children when the younger family members move away to find employment. This event may create the fragmentation of the family system by restricting ongoing social support and exposure to the norms of the traditional culture while affecting elders' social role and level of control within the family (Burr and Mutchler, 1993).

Beliefs and Values Towards Health

Hispanics tend to seek assistance for health-related problems from members within the group while Anglo-Americans seek help from more formal health care providers. For Hispanics, physical and mental health issues are intertwined reflecting traditional indigenous views of illness as involving mental, physical, and spiritual dimensions (Guarnaccia and Rodriguez, 1996; Rojas, 1997). This belief places control for one's health on supernatural causes, which are not in the immediate control of the individual. Many Mexican Americans seek the help of *curanderos*, who are native healers and who live in the immediate neighborhood. Cuban Americans also rely on spiritual practice (known as *Santeria*) to cope with health-related problems associated with stress caused by immigration, feelings of isolation, and loss of identity (Tran and Melcer, 2000). Unlike Anglo-Americans, Hispanics also strongly rely on the use of folk medicine as a substitute or complement to regular medical care (Kaiser, Gibbons, and Camp, 1993).

In sum, beliefs and values associated with health affect quality of life because elders' conceptualization of health carries implications for use of services, seeking support from family members, and perceived level of control over one's health.

Caregiving Practices and Attitudes

The role of family ties has an important influence on the quality of life of older people, especially related to their position within the family and the family's responsibility for caring for the elderly. The culture of ethnic minority groups, such as Hispanics is characterized by a traditional family structure formed of strong kin networks in which the nuclear family is embedded. The kinship network provides a source of psychological, emotional, and economic support for Hispanic elders. Cultural rules govern who provides care and how it is provided.

Traditional attitudes and circumscribed roles to women continue to be central to the Hispanic culture; women, especially adult daughters are expected to be the primary caregivers. For Hispanic elderly, the transition to the grandparent role is a very significant life-change event with very well defined grandparent roles and responsibilities. For Anglo-Americans, on the other hand, the responsibilities of becoming a grandparent may be less emphasized due to pressures of work, reduced number of extended family, and distance between children and parents. Hispanic elders also place great importance in the manner they are treated because of their age and need to assure their control and position in the family (Alemán, 2000).

The above-mentioned cultural factors impact the quality of life of immigrant elderly as they age in their adopted homeland. Their beliefs and values towards health and their caregiving practices influence how and where they wish to be cared for. However, because of poor and disparate access to health services, immigrant elderly will, at some point, need to be accommodated in the existing system of healthcare settings. At such a juncture, there is a need to examine the types of interventions that will offer immigrant elderly preferred attributes of quality of life. In the next section, we discuss the demonstrated effectiveness of a potential intervention and its impact on quality of life attributes.

Natural Environments as Settings Enabling Quality of Life Attributes–Engagement, Sense of Control, and Social Interaction

Research studies including a variety of settings and diversity of interactions with nature attest to people's preference for natural environments and their impact on psychological and physiological responses (Appleton, 1996; Kaplan and Kaplan, 1989; Ulrich, 1983; Hartig, Mang and Evans, 1991). Below, we describe studies that provide evidence of the beneficial effects of interaction with nature on engagement (decreased boredom), sense of control (decreased helplessness), and social interaction (decreased loneliness).

Use of Nature to Reduce Boredom and Facilitate Engagement

Effects of Boredom. Boredom is experienced when one is not free to do as one desires or when one feels that time is passing slower than expected. Among bored people, low internal arousal is associated with inattention, daydreaming, low performance, and sleepiness (Troutwine and O'Neal, 1981). An essential challenge of boredom therefore is the struggle to maintain attention. Institutionalized elders and those in health care environments are bored, at least in part, when they cannot keep their attention focused where it should be, or when they have no interesting stimuli to pay attention to.

Use of Nature to Facilitate Engagement

Research on the physiological effects of interaction with natural environments attests to its importance in buffering experiences of boredom. Scenes dominated by natural elements increase ratings of positive affect and reduce ratings of arousal among both stressed individuals (Ulrich, 1979) and unstressed individuals (Ulrich, 1981); moreover, interaction with nature leads to a shift towards a more positively-toned emotional state and positive changes in physiological activity levels (Ulrich, Dimberg, and Driver, 1991; Ulrich, 1981; Ulrich, 1984).

Activities such as gardening require continuous care, for it provides people with constant opportunities for nurturing and taking care of plants, creating a person-plant relationship that involves active participation (Lewis, 1996). For older adults in health care environments, natural environments can buffer experiences of boredom by providing sensory stimulation and opportunities to manipulate natural elements, to observe seasonal changes, and to actively engage in nature-related activities. Nature-related activities such as taking care of plants may also stimulate conversation and bring contact with other people when participating in horticultural activities, thereby counteracting boredom.

Use of Nature to Facilitate a Sense of Control

Effects of Lack of Control. Control has been defined as "the ability to regulate or influence intended outcomes through selective responding" (Rodin, 1986, p. 141). A closely related notion–helplessness (Seligman,

1975) or the failure to restore control, is the generalized perception that one's actions do not control one's outcomes. Aging lowers a perceived sense of control because elders undergo a series of physical and social changes that make them more vulnerable to a general loss of control due to diminished adaptive capabilities (Rodin and Timko, 1992). The experience of choice and control primarily impact the activities of daily living of elderly people that routinely support life, sustain competence and a sense of self (including maintaining continuity of self). Even a minimal amount of opportunities for choice over the performance of routine activities can be very significant for the general well-being of older adults (Langer and Rodin, 1976). The continuous use of rigidly programmed activities such as in an institutional healthcare setting that do not allow personal choice may instead lead to helplessness.

Use of Nature in Facilitating a Sense of Control

Natural environments provide an escape from current stressful psychological conditions that threaten one's sense of control. Such a temporary escape or being away, gives people the chance to distance themselves from stressful daily conditions and attempt to restore a sense of control. The notion of "thereness" (S. Kaplan, 1978) or the appreciation of nature (e.g., a park) by virtue of its availability, whether or not one participates in it has been identified as a significant contributor to a sense of control. People who "escape" negative psychological states achieve actual or perceived sense of control, which is the person's perceived ability to choose and determine what they do, to affect their life situations, and to determine what others do to them (Ulrich, 1999; Ulrich, Simons, Losito, Fiorito, Miles, and Zelson, 1991).

Another way nature provides control is by allowing people to change the focus of their attention from personal problems to something outside them, in the physical environment. Kaplan and Kaplan (1989)[1] called this process "soft fascination," which is achieved when the mind can wander when fixed on some aspect of the natural environment. Gardening experiences (R. Kaplan, 1973; S. Kaplan, 1978; Relf, 1992) elicit fascination by allowing people to function without having to pay too much attention to the environmental stimuli (such as flowers, leaves, and the elements part of a garden), thus, freeing their minds from stressful events.

Additionally, studies on urban gardening projects and tree-planting programs in low-income housing communities and schools demonstrate that when people work together to improve some aspect of the physical environment, they feel empowered and in control of what might happen in their communities (Lewis, 1996). Cooper Marcus and Barnes (1995) found that employees in healthcare facilities reported opportunities for escape from work stress and personal control as some of the benefits provided by gardens in hospital environments. McGuire (1997) found that control is elicited through nature-related activities that engage residents of nursing homes. Participation once a week in "garden club" activities conducted in a relaxed setting (e.g., selecting plants to be cared for and making decisions about materials and methods of planting) provided residents an opportunity to exercise their sense of control.

Use of Nature to Ameliorate Loneliness and Facilitate Social Interaction

Effects of Loneliness. Loneliness refers to negative psychological experiences associated with perceived social isolation or a lower level of social contact than what the individual desires (Penplau and Perlman, 1982). Advancing age and accompanying life-style changes may make people more vulnerable to experiences of loneliness which has been defined as "an aversive emotional experience that occurs when one's existing social relationships differ from one's desired social relationships" (Rook, 1984, p. 251). Loss of friends and family members, retirement, loss of cognitive and/or physical abilities to make and maintain social connections also contribute to feelings of loneliness (Wenger, Davies, Shahtahmasebi, and Scott, 1996). Peplau, Bikson, Rook and Goodchilds (1982) suggest that reduced feelings of loneliness among elderly people are enabled by availability of a confidant, personal control, and social comparison processes.

Cultural and ethnic factors also affect experiences of loneliness. Rathbone-McCuan and Hashimi (1982) identified a need to develop intervention programs that strengthen the social network of Hispanic elders in their communities. When they examined social isolation among Hispanic elderly living in the U.S., including Cubans, Puerto Ricans, and Mexican-Americans, they found that the demands of modern America have reduced some of the longstanding patterns of support for the elderly. The authors caution that the image of extended family caregiving in this community should not be widely generalized to overshadow the need for additional sources of support for the Hispanic elderly.

Use of Nature in Facilitating Social Interaction

Individuals who receive social support are less stressed and have a better health status than people who are socially isolated (Cohen and Syme, 1985). In seeking relevant intervention strategies to enhance the quality of life of institutionalized elders, some researchers have focused on the role of the natural environment as a way to increase social support and decrease loneliness. For example, gardens in healthcare institutions are considered important elements in fostering social and emotional support for patients, visitors, and employees (Ulrich, 1992a). Research with users of urban parks and other outdoor settings indicates that natural environments are used as settings for social and emotional support (Driver and Brown, 1986; Ulrich and Addoms, 1981).

There is limited information about the role of natural environments in fostering social support for elders in long-term care settings. However, Whall, Black, Groh, Yankou, Kupferschmid, and Foster (1997) showed that introducing bird songs, bird pictures, and the sound of babbling brooks was associated with decreased patient agitation and more positive affective responses among late dementia patients. Cohen and Day (1993) visited three Alzheimer's care facilities having gardens or planted courtyards in which the facility administrators noticed the preference of family members to visit with patients outdoors rather than indoors.

Leisure research (e.g., Hutchison, 1994) also points to the distinctive pattern of older people's activities in public parks–mostly stationary activities versus the sports-related activities of the younger group. Early in the morning, elders would gather together in previously established areas of the park to interact socially (e.g., talk to one another, play games, watch people). For them, the park served as a "neighborhood center" in facilitating interaction in small peer groups and supporting the maintenance of personal relationships.

Ethnicity influences patterns of preference and use of natural environments, including socially-oriented activities. Shaull and Gramann (1998) discovered that Hispanic Americans value outdoor recreational activities that promote social interaction and social bonds among their families. Since "familism" is one of the core cultural values for this group, the value of outdoor recreational activities lies in spending time with relatives to maintain social and cultural ties. Virden and Walker (1999) demonstrated that the use of public parks by Hispanics and Blacks is associated with the threat of other people because they view these settings as being more the domain of Anglo-Americans. This may lead groups to relate more to the members of their own groups and isolate themselves from interaction with other ethnic groups.

In general terms, African-American and Hispanic groups prefer higher levels of service development in natural settings than Anglo-Americans (Baas, Ewert, and Chavez, 1993; Dwyer and Hutchison, 1990), in part because the former are concerned with amenities that support visual access, allowing them to observe others and their activities, and evoking a consequent feeling of safety. Therefore, social interaction can develop only to the extent that people feel safe and welcome in a natural setting shared by groups from diverse ethnic backgrounds.

SUMMARY OF LITERATURE REVIEW

We focused on cultural differences between the elderly especially as it pertains to how each ethnic group might perceive the availability, access and use of health care resources. We also found that caregiving practices provided by conventional institutions conflict with core Hispanic caregiving values. Therefore, Hispanics do not seek institutionalization, preferring to stay with family instead. However, should the need arise, institutionalized Hispanic elders are more likely to feel like misfits within the conventional system of long-term care settings.

Our conceptual framework suggests that the common and major goal in settings that provide caregiving for the elderly should be to preserve the continuity of self. Therefore, we next examined the potential of environments, particularly nature settings as effective interventions in promoting continuity of self through experiences of engagement, control and social interaction. While there is evidence to suggest that all three are fostered by activities (such as gardening) in suitable natural environments, we found that there are potential differences in the way the same environments are perceived and used because of cultural patterns of social interaction and concerns for safety. As already emphasized, strong adherence to cohesive familism and filial obligations are some of the core issues of Hispanics cultural values and reflect in their choice for activities within outdoor and nature settings. Although nature settings have been influential in promoting social interaction, cultural differences in the use of nature may be critical factors influencing how we apply these findings to institutionalized elderly. Also, while nature is defined in most studies via its sensory qualities and individual elements such as trees, greenery, and shrubs, more information is needed on how these nature elements can be translated into a usable template for informing design. In our study we examine reactions by two cultural groups to six 'frames' of 'nearby nature' with varying degrees of proximity to built settings, enclosure, greenery and vegetation. In order to address some of the previously mentioned gaps in the literature review, this study used the following questions.

RESEARCH QUESTIONS ADDRESSED IN THIS STUDY

Familiarity and Ethnicity

The perception of control over one's environment is stronger when the demands of the situation are within one's competence. This notion of compatibility is easier to achieve in familiar environments, because familiarity brings with it some degree of predictability regarding what is expected in such an environment, and past experiences inform one's behavior and actions in that environment. Also, in the case of the elderly, recalling past life history contributes to a stronger continuity of self. Unfamiliar environments, on the other hand, potentially increase a sense of disorientation and helplessness. Therefore, to assess familiarity with the frames used in the study, we asked; *Do Hispanic and Anglo-American elderly differ in their familiarity with the nature frames?*

Preference and Ethnicity

Within predictable and familiar environments, people tend to prefer those that are potentially healthy and supportive based on preferred activities and experiences. In such environments, positive experiences are supported. For the elderly in particular, these preferred and positive activities and experiences might be important predictors of their quality of life and successful functioning in environments. Also, cultural factors such as preferred social interaction patterns might influence which environments are preferred. To find out more about environments preferred by the two ethnic groups, we asked; *Do Hispanic and Anglo-American elderly differ in their preference for the nature frames?*

Nature-Related Activities, Quality of Life Attributes and Ethnicity

Activities and experiences are the most enjoyable within familiar and preferred environments, and contribute to a sense of well-being. As the literature review indicates, nature-related activities foster engagement, social support and a sense of control–all positive reinforcements of increased quality of life. The literature also indicated that cultural differences might be evident in the nature-related activities. Therefore, we asked; *Do Hispanic and Anglo-American elderly differ in the kinds of activities and experiences they associate with the nature frames? Which attributes of quality of life are associated with these activities and experiences?*

METHODOLOGY

Sample

The sample of participants included 15 Spanish-speaking Hispanic and 15 English-speaking non-Hispanic Anglo-American elderly people. The Hispanic elderly were members of an adult day care center in Wisconsin and the non-Hispanic Anglo-American elderly were members of a suburban independent living retirement community in Illinois. The elderly people were recruited with the help of the Program Directors at both places.

The sample of nature photographs chosen for this study included six categories; courtyard, gazebo (see Figures 3-5), green lawn, pond, patio and plaza (see Figure 6). The picture of the pond showed a lily pond in the foreground surrounded by low lying bushes. The picture of the patio showing a brick patio roofed by trellis. Four chairs surrounded a table in the center of the patio, while potted plants flank its sides. We chose pictures that reflected a range of possibilities for engagement, social interaction and control. These pictures reflect different levels of human intervention represented by the presence of human artifacts–two categories, green lawn and pond are the most 'natural' with very few human artifacts whereas the other four categories show a greater presence of human artifacts.

Instruments, Procedure and Data Analysis

All 30 participants were interviewed using a semi-structured interview guide. The six color printed frames, 8 1/2" × 11" each, were shown one by one during the course of the interview. For each frame, the questions focused on familiarity with the nature setting represented in the nature frame, the types of previously experienced/projected activities engaged by the participant in that place and with whom, and whether any other memory could be associated with that place. At the end of the first part of the interview, the frames were placed on a table in the same

FIGURE 3. Courtyard

Photo Courtesy: Karen S. R. Balogh. In Gerlach-Spriggs, Kaufmen, & Warner (1998) *Restorative Gardens. The Healing Landscape*. Page 60. New Haven: Yale University Press.

FIGURE 4. Gazebo

Photo Courtesy: Richard Payne. In Gerlach-Spriggs, Kaufmen, & Warner (1998) *Restorative Gardens. The Healing Landscape*. Page 93. New Haven: Yale University Press.

FIGURE 5. Green Lawn

Photo Courtesy: Uriel Cohen

order they were presented. Next, the participants were asked to rank the photographs based on preferred experienced/projected activities in each nature frame.

For the non-Hispanic Anglo-American elderly, each interview was conducted at a place chosen by the residents, some were in their apartments, but most were conducted in the private dining room at the facility. For the Hispanic elderly, the interpreter conducted each interview in Spanish at the adult day care center. The responses were translated back into English to enable a comparative data analysis. The data obtained from the open-ended questions were coded independently by two researchers and consensus reached.

RESULTS

Familiarity and Ethnicity

Do Hispanic and Anglo-American elderly differ in their familiarity with the nature frames?

The respondents were asked whether they had been to a place like the one shown in the frame or if they were familiar with such a place

FIGURE 6. Plaza

Photo Courtesy: David Hewitt Anne Garrison Photos. In Donald J. Canty (1990) Urban Delight. Page 65. *Architectural Record* October 1990.

(see Figure 7). The Anglo-American elderly were most familiar with courtyard and green lawn categories, although a high percentage in this group was familiar with all the frames (gazebo-87%, pond-80%, patio-73%, and plaza-87%). The Hispanic elderly were most familiar with the green lawn (93%), plaza (80%) and patio (73%) categories and least familiar with the gazebo (60%) and pond (60%) categories.

Preference and Ethnicity

Do Hispanic and Anglo-American elderly differ in their preference for the nature frames?

All the respondents were asked to rank the frames based on their preferred and most liked activities in each place (see Figures 8 and

FIGURE 7. Familiarity with the Frames (Shown as Stimuli) by Ethnicity

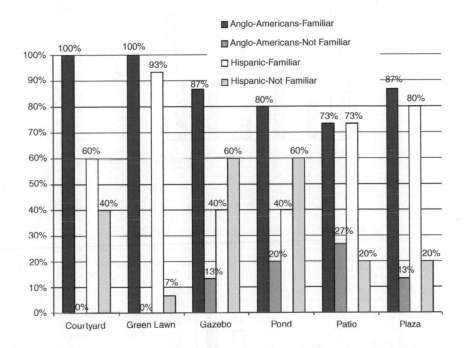

9). While 47% of the Hispanic elderly preferred the courtyard and 40% the plaza the most, the Anglo-American elderly were equally divided in their preferences for courtyard (20%), gazebo (20%), patio (20%) and plaza (20%). The Hispanic elderly least preferred the patio (40%) and pond (26.7%), and the Anglo-American elderly least preferred the plaza (46.7%) and patio (26.7%).

Nature-Related Activities, Quality of Life Attributes and Ethnicity

Do Hispanic and Anglo-American elderly differ in the kinds of activities and experiences they associate with the nature frames? Which quality of life attributes are associated with these activities and experiences?

For this question, the open-ended responses by both ethnic groups were coded independently by two researchers based on the definitions

FIGURE 8. Most Preferred Frame by Ethnicity

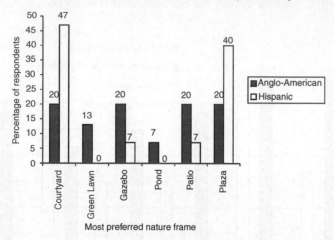

Most preferred nature frame

FIGURE 9. Least Preferred Frame by Ethnicity

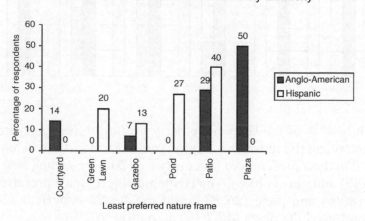

Least preferred nature frame

for engagement, social interaction and a sense of control. Engagement referred to the person making mention of passive (gazing, watching, enjoying looking at scenery, flowers, etc.) or active (swimming, walking through, playing) involvement with the setting. Social interaction was operationally defined based on the presence of any other person/s involved in the activity with her/him. Sense of control referred to the acknowledgement of a nature-related activity that reinforces a sense of control, one's ability to perform tasks and take care of nature-related elements (e.g., "I pruned the roses there," "I would like to talk to the plants, to water them, and take care of them"). The following analysis presents the responses according to: (a) activities associated with each frame and (b) attributes of quality of life implied in the responses.

Analysis in Terms of Activities Associated with Each of the Nature Frames

Courtyard. For Hispanic elderly, activities are described in terms of social interaction (e.g., I would like to play, talk, and share time with friends or people), exercising control (e.g., I would like to talk to the plants, water them, and take care of them), and passive engagement (e.g., I would like to observe nature and think how beautiful life is). For Anglo-American elderly, the courtyard is described in terms of engagement and social interaction. Activities indicate a high level of both active and passive engagement (mentioned by 80% of the respondents.) Walking and watching/enjoying the flowers were mentioned by most; some responses indicated exercising control (e.g., "I could be weeding, cutting flowers here . . ."). The courtyard was also a setting to be enjoyed with others.

Green Lawn. For Hispanics, the green lawn is described in terms of "active" engagement. Activities are related to practicing sports (e.g., I would like to play some sports, run or just exercise a bit) and indicate a high level of engagement (but not necessarily with others). For Anglo-American elderly, the green lawn also entails engagement. Activities recalled by the respondents indicate active sports (e.g., "canoeing," "playing baseball") and passive pursuits (e.g., "fishing," "sit down and enjoy the water"). While most activities almost always involved other people (e.g., picnics) other activities were solitary (e.g., "a place to be in silence and to be by yourself a bit").

Gazebo. For Hispanics, the gazebo is associated with a high level of "passive" engagement (related to relaxation, reflection, and fascination) and social interaction. Activities are oriented towards relaxation/reflection (e.g., watch the flowers and enjoy nature) and social interaction (e.g., I would like to chat with friends and have a good time with them). For Anglo-Americans, the gazebo is described in terms of "passive" engagement and some social interaction. Activities involved watching the azaleas in bloom (e.g., "I would go here when everything is in bloom," "when you see something that pretty, you just have to sit down and look around and enjoy it"), and socializing with family and friends (e.g., "a lovely place for a picnic or a family gathering").

Pond. For Hispanics, the pond brings experiences of "passive" engagement such as enjoying nature; listening to the birds and enjoying the fresh air during the morning. For Anglo-American elderly, the pond also associated with "passive" engagement. Activities were mostly centered around quiet engagement (e.g., "the lily pads are very pretty") although other activities like walking and fishing are also mentioned.

Patio. For Hispanics, the patio was associated with experiences of social interaction (e.g., I would like to play dominoes and share time with family and friends) and sense of control (e.g., I would like to cut the grass every six weeks). Likewise, Anglo-Americans associated the patio with experiences of social interaction. All the activities were centered on social interaction or enjoying the patio for dinner, reading and entertaining.

Plaza. For Hispanics, the plaza was associated with social interaction and engagement. Activities reflect both a high level of engagement (e.g., I would take a walk and watch people going back and forth) and social interaction (e.g., I would like to sit down and chat with friends). For Anglo-American elderly, the plaza was associated mostly with social interaction. Activities were mostly related to walking through the space and shopping.

Analysis in Terms of Quality of Life Attributes Implied in the Responses

Social Interaction. Hispanic elderly preferred patio and plaza for activities that facilitated social interaction whereas Anglo-American elderly preferred gazebo and patio for social interaction. The patio seems to be a preferred setting for the realization of social interaction for both groups. Hispanics mentioned activities that reflected interaction with others, such as playing dominoes and sharing time with family and friends. Anglo-American elderly mentioned activities that centered on social interaction or enjoying the patio for dinner, reading and entertaining.

Sense of Control. For Hispanic elderly, the plaza was the setting most associated with sense of control whereas for Anglo-American elderly it was the courtyard. For Hispanics, sense of control was expressed in activities such as cutting the grass, having birds and plants, and having a garden with roses. For Anglo-American elderly, control was reflected in answers such as weeding and cutting flowers.

Engagement. The two groups differed in their choice of settings for the realization of engagement. Hispanics preferred the pond for activities that facilitated engagement whereas Anglo-American elderly preferred the courtyard. For Hispanics, the pond provided the realization of passive engagement or activities that allowed one to relax and enjoy nature. Anglo-American elderly described the courtyard as facilitating passive engagement, such as relaxing and enjoying nature. In this case, the kind of engagement provided by the different setting was of the same nature.

Summary of Nature-Related Activities, Quality of Life Attributes and Ethnicity

Overall, there are some similarities and differences among the activities and experiences associated with natural environments by the two ethnic groups.

A striking difference is the absence of green lawn and pond–the most natural of the categories, from the Hispanics most preferred list. In fact, these two categories appear in the least preferred list (see Figure 9). The green lawn, for example, is very familiar to the Hispanics (to 93% of the respondents) but is least preferred by the respondents as settings for their favored activities. Despite being fairly high in engagement (see Figure 10), the lawn is lower in terms of social interaction and providing a sense of control. Similarly the pond, while highest in terms of engagement (73% of Hispanic respondents), is lower in terms of a sense of control (20%) and the least (7%) in terms of social interaction and is therefore preferred the least by 26.7% of Hispanics.

On the other hand, the plaza (87% of Hispanics respondents were very familiar and 40% preferred it the most) and the courtyard (preferred the most by 46.7%) are most conducive to social interaction– 47% and 33% of Hispanics respectively mention social affiliation in these places. Additionally, the courtyard also offers opportunities to exercise control (27%). The patio, however–which is familiar (73%) is high in terms of social interaction (47%) and a sense of control (40%) but is least preferred by most (40%) of the Hispanics. This is perhaps due to the low engagement–only 13% of the respondents report any type of engagement in the patio.

For the Anglo-American elderly, high familiarity with almost all the settings is reflected in the most preferred list–they are equally divided in terms of which frame they liked the most for the pursuit of activities. However, the least preferred setting is the plaza (50%) which was considered familiar by 87% of the respondents. Figure 10 reveals that for the Anglo-American elderly, the plaza is relatively low on engagement

FIGURE 10. Quality of Life Attributes Related to Each Frame

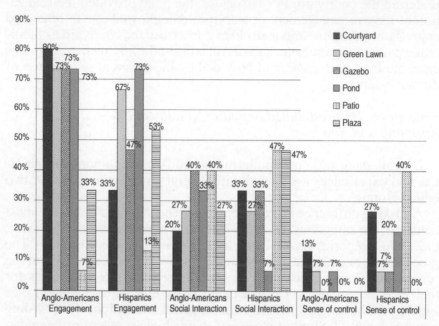

(33%) when compared to other settings, low on social interaction (27%) and low in exercising control (0%). Similarly, the patio–least preferred by 29% of the Anglo-American elderly, is lowest in terms of engagement (7%) and exercising control (0%) although it does provide social interaction (40%).

Therefore, it appears that Hispanic elderly prefer natural settings such as the courtyard and the plaza because they facilitate social interaction and the realization of activities with others. Anglo-American elderly seem to be less concerned about the choice of settings for the realization of activities. For them four of the six frames were equally preferred for the pursuit of activities.

RETROSPECT AND IMPLICATIONS

Our conceptual framework included the sense of control as an attribute of quality of life of elderly. The perception of control over one's environment is stronger and easier to achieve in familiar environments, because familiarity brings with it some degree of predictability regarding what is expected in such an environment. Unfamiliar environments, on the other hand, potentially increase a sense of disorientation and helplessness. Most Anglo-American elderly were familiar with all the frames, expressing the most familiarity with the courtyard and green lawn. This is not surprising because the types of nature frames chosen for this exploratory study are highly prevalent in the United States. Most Hispanic elderly were familiar with the green lawn, patio, courtyard, and plaza. The Hispanic elderly also related the frames they were familiar with to their past experiences in Mexico, for example. The Hispanic elderly respondents were most unfamiliar with the gazebo and pond categories. These types of nature settings might not have been a major part of their past life and residential history outside the United States.

We surmised earlier that within familiar environments people tend to prefer those that are potentially healthy and supportive based on preferred activities and experiences. For the elderly in particular, preferred and positive culturally comfortable activities, interaction and experiences might be important predictors of attributes of quality of life such as social interaction and engagement. Based on the favored activities in the places, the Hispanics preferred the courtyard and the plaza the most. Among the least preferred categories for the Hispanic respondents were the gazebo, pond, green lawn and patio. The Anglo-American elderly were equally divided among the nature frames in terms of preferred activities.

Although limited by the relatively small sample of nature frames and people interviewed, this study opens the door for further focused exploration of culturally influenced nature-related activities among elderly. It is significant because it has identified that there are differences in the way the same nature settings are perceived by two different ethnic elderly groups of potential users.

We began the study with an assumption that one size may not fit all, especially when it comes to nature interventions to ameliorate a common set of problems for the institutionalized elderly–loneliness, boredom and helplessness. This assumption seems to hold true for at least these two ethnic groups–different nature settings are preferred by each group based on different sets of preferred activities in each setting. What does this indicate to future researchers and future designers?

In terms of specific interventions, we urge researchers to include a conceptualization of cultural factors when mapping relationships between environments, users and activity planning. The cultural context may offer *activity-motivating* factors that serve as a trigger or a catalyst to action, *performance-enhancing* factors contributing to involvement, engagement, and sustained interest, and activities that may have an *enhanced affect*, such as affirmation, satisfaction, personal meaning and improved self image (Cohen & Cohen, forthcoming, 2005).

There are many ways to conceptualize the elderly person, the environment, and beneficial interactions. We chose the notion of quality of life as central to our understanding of the elderly person's living world. Other conceptualizations will benefit greatly as well by including cultural factors. From our exploratory study, it is clear that there are differences in the nature-related activities associated with each setting by the two ethnic groups. There are also differences among the groups in terms of the engagement, social interaction and sense of control enabled in each setting. In fact, the sense of control was the most elusive of the quality of life attributes, and further research is much required to elaborate this relationship in specific terms.

In the design realm, we specifically looked at ways in which nature could be potentially integrated with long term care settings. Based on the findings, two types of nature settings seem to take shape based on the preferred activities mentioned by the respondents:

1. 'furnished' nature settings such as the plaza which has elements like informal seating, amenable to activities such as social interaction, and
2. 'authentic' nature settings such as the pond which is more amenable to quiet reflection and soft fascination as the Kaplans describe.

The findings also show that Hispanic respondents are drawn to the furnished nature settings more readily and comfortably than the Anglo-American respondents who are drawn more towards the authentic nature settings.

SUGGESTED DESIGN RESPONSES

Based on the above important observations, we propose the following ways in which nature can be integrated in the design of health care and long term care settings for the multicultural elderly:

Flexible and Intimate Spaces for Different Levels of Engagement and Social Interaction

We notice in this study that nature settings are appreciated by the respondents for their intrinsic qualities and passive engagement such as aesthetic beauty and tranquility, and for the extrinsic qualities such as activities they enable, for example, walking though the garden, finding a place to sit and enjoy the view, ability to get one's feet in the water, etc.

Suggested Design Responses:

- A balanced approach to provide 'furnishings' such as built-in and moveable furniture while also retaining the 'authentic' natural elements such as trees, grass, water feature, and flowers.
- Providing a vista with informal seating such as benches or ledges when possible so that authentic nature settings can be viewed comfortably.
- Enabling access to nature, for example, providing an unhindered pathway towards a pond or lake, providing opportunities to touch the water such as a bird bath or a small fountain.

Variety of Nature Settings to Allow for Individual Choice of Activities

The study revealed that the two ethnic groups vary in their choice of activities within the same nature settings. Some settings allowed for solitary activities by both ethnic groups, for example, the pond. Even when the presence of other people is mentioned, there is a difference in the type of interaction. To some Anglo-American respondents, the presence of a loved one enhances the nature experience (that is, watching the birds along with a spouse); to Hispanic respondents, more 'active' interaction such as talking is preferred.

Suggested Design Responses:

- Providing well separated functional nature enclosures or outdoor 'rooms' so that different types of quiet and 'active' activities can occur simultaneously.
- Providing nature areas that enable both solitary and social activities.

- Providing buffers such as tree-lined paths or hedges with clear visual access.

Open-Ended Design to Allow Elderly Users to Change and Nurture the Environment

The study revealed that active engagement with nature, such as planting and watering flowers, weeding, etc., are mentioned by respondents from both ethnic groups. Other activities include tending to pets such as fish and birds in a nature setting (mentioned mostly by Hispanic respondents) and taking pictures of nature.

Suggested Design Responses:

- Taking precautions not to over-design the nature settings, especially if it is a new long-term facility, instead, providing multiple areas that can grow along with the residents, in the form of small gardens, ponds with fish and flowers, etc.
- Integrating design elements such as bird baths and bird feeders can potentially engage at least two types of users; those who are interested in tending to the birds, and those who are interested in watching or photographing the birds.

Organizational Support to 'Personalize' the Nature Setting to Make the Setting More Home-Like and Provide Continuity of Self

While our study asked the respondents to share potential use of nature settings, we did not focus on the organizational aspect. However, as in any design venture, the organizational responsibility must be addressed so that settings are optimally used. Particularly while designing nature-related activities such as horticultural therapy, it is important to realize that cultural differences may influence success.

Suggested Design Responses:

- To enable meaningful nature interactions, activities can be designed to suit multiple interaction levels, for example, providing a choice of either gardening, working with nature elements to make crafts, or photographing or painting nature.
- A family-centered philosophy that engages friends and family in frequent nature-related activities such as picnics in the park, trea-

sure hunts, outdoor music concerts, hiking, and fishing activities. These activities can be designed to be either intimate or accommodate larger numbers of people as the need may be.

CONCLUSION

This study provides a better understanding of the role of natural elements as health-promoting tools in long-term and health care settings, and heightens an awareness of how nature interactions might offer an inexpensive alternative/supplement to traditional medical treatment. Particularly, by emphasizing the role of cultural factors in this beneficial relationship between elderly and nature, it strengthens future design efforts for multicultural health care settings. While much work is warranted both in terms of research into other cultural groups, and testing and refinement of designed nature settings, this study is a first and important step.

NOTE

1. Kaplan and Kaplan identify four properties of a restorative experience (that are not unique to natural settings). They argue that nature benefits people suffering from mental fatigue because it offers the opportunity to escape (being away) as well as the opportunity to perceive the environment as part of a larger whole (extent); it allows people to function without considerable mental effort (fascination); and it provides satisfaction to people when it corresponds to their inclination and actions in the environment (compatibility).

REFERENCES

Alemán, S. (2000). Mexican-American Elders. In S. Alemán, T. Fitzpatrick, T. V. Tran, and E. Gonzales (Eds.), *Therapeutic Interventions with Ethnic Elders: Health and Social Issues.* New York: The Haworth Press.

Angel, J. L. (1991). *Health and Living Arrangements of the Elderly.* New York: Garland.

Angel, J. L., Angel, R. J., and Markides, K. S. (1996). Nativity, declining health and preferences in living arrangements among elderly Mexican Americans: Implications for long-term care. *The Gerontologist, 36*, 464-473.

Appleton, J. (1996). *The Experience of Landscape.* New York: John Wiley & Sons.

Baas, J. M., Ewert, A. W., & Chavez, D. J. (1993). Influence of ethnicity on recreation and natural environment use patterns: Managing recreation sites for ethnic and racial diversity. *Environmental Management, 17*, 523-529.

Burr, J. A., and Mutchler, J. E. (1993). Nativity, acculturation, and economic status: Explanations of Asian American living arrangements in later life. *Journals of Gerontology: Social Sciences, 48*, 555-563.

Cohen, S. and Syme, S. L. (Eds.). (1985). *Social Support and Health.* New York: Academic Press.

Cohen, U. and Cohen, R. Environment, culture, and physical activity for older persons. In Wiemo, Z. (Ed.), *Measurement and Activity of Older Persons.* American College of Sports Medicine (Forthcoming, 2005).

Cohen, U. and Day, K. (1993). *Contemporary Environments for People with Dementia.* Baltimore: John Hopkins University Press.

Cooper Marcus, C. and Barnes, M. (1995). *Gardens in Healthcare Facilities: Uses, Therapeutic Benefits, and Design Recommendations.* Martinez, CA: The Center for Health Design.

Cox, C. (1993). *The Frail Elderly: Problems, Needs, and Community Responses.* Connecticut: Auburn House.

Day, K., and Cohen, U. (2000). The role of culture in designing environments for people with dementia: A study of Russian Jewish immigrants. *Environment and Behavior, 32*(3), 361-399.

Driver, B. L., and Brown, P. J. (1986). Probable personal benefits of outdoor recreation. In *President's Commission on Americans Outdoors: A Literature Review.* Washington, DC: Government Printing Office, 63-67.

Dwyer, J. F., and Hutchinson, R. (1990). Outdoor recreation participation and preferences by Black and Anglo-American Chicago households. In J. Vining (Ed.) *Social science and natural resource management.* Boulder, CO: Westview Press.

Eribes, R. A., and Bradley-Rawls, M. (1978). The underutilization of nursing home facilities by Mexican-American elderly in the southwest. *The Gerontologist, 18*, 363-371.

Godbey, G., and Blazey, M. (1983). Old people in parks: An exploratory investigation. *Journal of Leisure Research, 15*, 229-244.

Guarnaccia, P. J., and Rodriguez, O. (1996). Concepts of culture and their role in the development of culturally competent mental health services. *Hispanic Journal of Behavioral Sciences, 18*, 419-443.

Hartig, T., Mang, M., and Evans, G. W. (1991). Restorative effects of natural environment experiences. *Environment and Behavior, 23*, 3-36.

Hutchison, R. (1994). Women and the elderly in Chicago's public parks. *Leisure Sciences. 16*, 229-247.

Kaiser, M. A., Gibbons, J. E., and Camp, H. J. (1993). Long-term care: Development of services for Latino elderly. In M. Sotomayor and A. Garcia (Eds.), *Elderly Latinos: Issues and Solutions for the 21st Century.* Washington, DC: National Hispanic Council on Aging.

Kaplan, R. (1973). Some psychological benefits of gardening. *Environment and Behavior, 5*, 145-161.

Kaplan, R. and Kaplan, S. (1989). *The Experience of Nature: A Psychological Perspective.* Cambridge: Cambridge University Press.

Kaplan, S. (1978). Attention and fascination: The search for cognitive clarity. In S. Kaplan & R. Kaplan, Eds. *Humanscape: Environments for People.* (pp. 84-90). Belmont, CA: Duxbury (Ann Arbor, MI: Ulrich's Books, 1982).

Kessler, R. C., and Clearly, P. D. (1980). Social class and psychological distress. *American Sociological*, *45*, 463-478.

Langer, E. J., and Rodin, J. (1976). The effects of choice and enhanced personal responsibility for the aged: A field experiment in an institutional setting. *Journal of Personality and Social Psychology*, *34*(2), 191-198.

Lawton, M. P. (1991). A multidimensional view of quality of life in frail elders. In J. E. Birren, J. Lubben, J. Rowe, and D. Deutchman (Eds.), *The Concept and Measurement of Quality of Life in the Frail Elderly*. San Diego: Academic Press.

Lewis, C. A. (1996). *Green Nature/Human Nature: The Meaning of Plants in our Lives*. Urbana, IL: University of Illinois Press.

Miranda, M. R. (1991). Mental health services and the hispanic elderly. In M. Sotomayor, *Empowering Hispanic Elderly Families: A Critical Issue for the 90s*. Milwaukee, WI: Family Service America.

Peplau, L. A., Bikson, T. K., Rook, K. S., and Goodchilds, J. D. (1982). Being old and living alone. In L. A. Peplau & D. Perlman (Eds.) *Loneliness: A Sourcebook of Current Theory, Research and Therapy*. New York: John Wiley.

Peplau, L. A., and Perlman, D. (1982). *Loneliness: A Sourcebook of Current Theory, Research, and Therapy*. New York: Wiley.

Rathbone-McCuan, E. and Hashimi, J. (1982). *Isolated Elders: Health and Social Intervention*. Rockville, Maryland: Aspen Publications.

Relf, D. (1992). The *Role of Horticulture in Human Well-Being and Social Development*. Portland, OR: Timber Press, Inc.

Rodin, J. (1986). Health, control, and aging. In M. M. Baltes, and P. B. Baltes (Rds.), *The Psychology of Control and Aging*. Hillsdale, NJ: Lawrence Erlbaum Associates.

Rodin, J. and Timko, C. (1992). Sense of control, aging, and health. In M. G. Ory, R. P. Abeles & P. D. Lipman (Eds.), *Aging, Health, & Behavior*. Newbury Park, CA: Sage Publications.

Rojas, D. Z. (1997). Spiritual well-being and its influence on the holistic health of Hispanic women. In S. Torres (Ed.), *Hispanic Voices: Hispanic Health Educators Speak Out*. New York: NLN Press.

Rook, K. S. (1984). Research on social support, loneliness, and social isolation: Toward an integration. *Review of Personality and Social Psychology*, *5*, 239-264.

Seligman, M. E. P. (1975). *Helplessness: On Depression, Development and Death*. San Francisco, CA: Freeman.

Shaull, S. L., and Gramann, J. H. (1998). The effect of cultural assimilation on the importance of family and nature-related recreation among Hispanic Americans. *Journal of Leisure Research*, *30*(1), 47-63.

Sotomayor, M., and Curiel, H. (1988). *Hispanic Elderly: A Cultural Signature*. Edinburg, TX: Pan American University Press.

Tran, T. V., and Melcer, N. C. (2000). Cuban-American Elders. In S. Alemán, T. Fitzpatrick, T. V. Tran, and E. Gonzales (Eds.), *Therapeutic Interventions with Ethnic Elders: Health and Social Issues*. New York: The Haworth Press.

Troutwine, R. and O'Neal, E. C. (1981). Volition, performance of a boring task and time estimation. *Perceptual and Motor Skills*. *52*(3), 865-6.

Ulrich, R. S. (1979). Visual landscapes and psychological well-being. *Landscape Research*, *4*(1), 17-23.

Ulrich, R. S. (1981). Natural versus urban scenes: Some psychophysiological effects. *Environment and Behavior, 13*, 523-556.

Ulrich, R. S. (1983). Aesthetic and Affective Response to Natural Environment. In I. Altman and J. F. Wohlwill (Eds.), *Human Behavior and Environment, Vol. 6: Behavior and the Natural Environment.* New York: Plenum Press.

Ulrich, R. S. (1984). View through a window may influence recovery from surgery. *Science, 224,* 420-421.

Ulrich, R. S. (1992). Effects of interior design on wellness: Theory and recent scientific research. *Journal of Healthcare Design, 3,* 97-109.

Ulrich, R. S. (1999). Effects of gardens on health outcomes: Theory and research. In C. C. Marcus and M. Barnes (Eds.), *Healing Gardens: Therapeutic Benefits and Design Recommendations.* New York: John Wiley & Sons.

Ulrich, R. S. and Addoms, D. L. (1981). Psychological and recreational benefits of a residential park. *Journal of Leisure Research, 13,* 43-65.

Ulrich, R. S., Dimberg, S. U., Driver, B. L. (1991). Psychophysiological indicators of leisure benefits. In B. L. Driver, P. J. Brown, and G. L. Peterson (Eds.), *Benefits of Leisure.* State College PA: Venture Publishing.

Ulrich, R. S., Simons, R. F., Losito, B. D., Fiorito, E., Miles, M. A., and Zelson, M. (1991). Stress recovery during exposure to natural and urban environments. *Journal of Environmental Psychology. 11*(3), 201-230.

U. S. Bureau of the Census. (2000). Census 2000 Summary File 1(SF 1) 100-Percent Data.

Virden, R. J. & Walker, G. J. (1999). Ethnic/racial and gender variations among meanings given to, and preferences for, the natural environment. *Leisure Sciences, 21,* 219-239.

Wenger, G. C.; Davies, R.; Shahtahmasebi, S., and Scott, A. (1996). Social isolation and loneliness in old age: Review and model refinement. *Aging and Society, 16,* 333-358.

Whall, A. L., Black, M. E., Groh, C. J., Yankou, D. J., Kupferschmid, B. J., and Foster, N. L. (1997). The effect of natural environments upon agitation and aggression in late stage dementia patients. *American Journal of Alzheimer's Disease, 12,* 216-320.

Presence and Visibility of Outdoor and Indoor Physical Activity Features and Participation in Physical Activity Among Older Adults in Retirement Communities

Anjali Joseph
Craig Zimring
Lauren Harris-Kojetin
Kristen Kiefer

SUMMARY. In this paper we examine how the presence and visibility of outdoor and indoor physical activity resources (e.g., walking path/trail, outdoor tennis courts, gardens, etc.) influences participation in physical activity among elderly residents in non-profit continuing care retirement communities and other senior housing communities. This pa-

---

Anjali Joseph, PhD, is Director of Research at the Center for Health Design, based in Concord, California (E-mail: ajoseph@healthdesign.org). Craig Zimring, PhD, is affiliated with the Georgia Institute of Technology. Lauren Harris-Kojetin, PhD, is affiliated with the Institute for Future of Aging Services, American Association of Homes and Services for the Aging. Kristen Kiefer, MPP, is affiliated with the Institute for Future of Aging Services, American Association of Homes and Services for the Aging.

This project was funded by a grant from the Robert Wood Johnson Foundation (RWJF).

[Haworth co-indexing entry note]: "Presence and Visibility of Outdoor and Indoor Physical Activity Features and Participation in Physical Activity Among Older Adults in Retirement Communities." Joseph, Anjali et al. Co-published simultaneously in *Journal of Housing for the Elderly* (The Haworth Press, Inc.) Vol. 19, No. 3/4, 2005, pp. 141-165; and: *The Role of the Outdoors in Residential Environments for Aging* (ed: Susan Rodiek, and Benyamin Schwarz) The Haworth Press, Inc., 2005, pp. 141-165. Single or multiple copies of this article are available for a fee from The Haworth Document Delivery Service [1-800-HAWORTH, 9:00 a.m. - 5:00 p.m. (EST). E-mail address: docdelivery@haworthpress.com].

Available online at http://www.haworthpress.com/web/JHE
© 2005 by The Haworth Press, Inc. All rights reserved.
doi:10.1300/J081v19n03_08

per reports findings from a survey of 800 such communities. A social ecological model was used to study the relationships between the environment and physical activity behavior. A fifty-two percent response rate (n = 398) was obtained. Campuses with more attractive outdoor and indoor physical activity facilities had more residents participating in different types of physical activity. *[Article copies available for a fee from The Haworth Document Delivery Service: 1-800-HAWORTH. E-mail address: <docdelivery@haworthpress.com> Website: <http://www.HaworthPress.com> © 2005 by The Haworth Press, Inc. All rights reserved.]*

KEYWORDS. Physical activity, older adults, outdoor features

INTRODUCTION

Regular physical activity contributes to better health among old and very old individuals, allowing them to remain independent for a longer period of time (Shephard, 1997). However, despite the well-established benefits of routine moderate physical activity for older adults, this segment of the U.S. population is the most sedentary, with inactivity being particularly pervasive among people 75 and older (King, Rejeski, & Buchner, 1998; USDHHS, 1996). As with other populations, public health policymakers and researchers are increasingly examining the role of the physical setting in encouraging or discouraging physical activity. For example, one of the strategies identified in the National Blueprint on Physical Activity Among Adults Age 50 and Older to enhance health and increase physical activity among older adults is "to create, promote and sustain communities that support lifelong physical activity" including physical settings that support activity (RWJF, 2000. p. 28).

Researchers from different fields such as public health, recreation science, urban planning and architecture are providing convergent evidence that neighborhood design is associated with physical activity by older people. For example, factors shown to encourage older adults to be active include the presence of walkable green areas and tree-lined walking paths near residence (Takano, Nakamura, & Watanabe, 2002), aesthetic beauty of the neighborhood (Brownson et al., 2000), safe and well-maintained walking paths in the neighborhood (Booth, Owen, Bauman, Clavisi, & Leslie, 2000) and convenient location and access to exercise facilities (Booth et al., 2000; Brownson et al., 2000; Carnegie, 2002).

While many older adults spend the vast majority of their day in and around buildings, there is much less rigorous research focusing on the impact of design features at the spatial scale of the site, campus or building. Recommendations from case studies and observations at residential communities for older adults suggest that visibility of exercise related areas from public and semi-public areas within a building (Howell, 1980; Regnier, 1994), views to the outside from exercise rooms, presence of walkable spaces within the facility (Regnier, 1994), perceived safety of outdoor spaces as well as the presence of interesting destinations within the facility (Parker and Joseph, 2003) may be factors that encourage older adults to be active. Further, it is plausible that factors such as availability of resources for physical activity, that have been shown to influence participation in physical activity at the neighborhood scale, may also be linked to physical activity behaviors at building and site scales.

Most previous research on the impact of the environment on physical activity behavior has focused on older people who live in apartments and homes in the community. An estimated 600,000 Americans live in continuing care retirement communities (CCRCs) and other retirement facilities. This number is rapidly increasing as the baby boom generation ages (AAHSA, 2005).

This questionnaire study examines what environmental support for physical activity is available at the site and building scale of continuing care retirement communities (CCRCs) and how the presence of physical activity outdoor and indoor features is related to self-reported physical activity. In the following sections we provide a model for organizing the research, define key terms and describe the research methods used. This is followed by discussion of the results of the survey and implications for future work.

Theoretical Concerns

The physical environment interacts with a host of other factors in influencing an individual's decision to be physically active. We adopted a *social ecological model* for this study that acknowledges the multiple factors that influence an older person's decision to be active. Social ecology models seek to understand complex patterns of causation where individual and group behaviors are influenced by, and influence, social and physical structures (Satariano & McAuley, 2003; Zimring, Joseph, Nicoll & Tsepas, 2005). As illustrated in Figure 1, we see physical activity as related to environmental factors, but where organizational and personal factors both moderate the role of the environment and have direct effects. *Personal Factors* include demographic and

FIGURE 1. A Social Ecological Model of Influences on Physical Activity

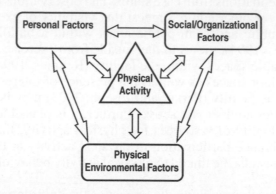

Source: Zimring, Joseph, Nicoll, Tsepas (2005). (Kerr, Eves, & Carroll, 2001)

health variables, an individual's knowledge, attitudes and beliefs related to physical activity and psychological or behavioral attributes and skills that may facilitate or impede efforts to participate in physical activity (King, 2001). Age is an important factor influencing participation in physical activity. In a survey of cognitively intact subjects aged 90 and older it was found that age was negatively related to physical activity (Hilleras et al., 1999). *Social/Organizational Factors* include the goals, philosophies and culture of organizations and social structure and support which may facilitate or impede efforts to participate in physical activity. This includes the type and number of physical activity programs that are easily available to older adults (King, 2001). *Physical Environmental Factors* can be considered at four nested levels of spatial scale: (1) urban design; (2) site selection and design; (3) building design; and, (4) building element design.

The physical environment offers different resources and constraints to participation in physical activity at different spatial scales. For example, issues such as traffic safety and land use mix may be important factors affecting participation in physical activity at the urban scale while factors such as location of social areas and views to interesting destinations may be important for walking within buildings. Most of the physical activity-environment research is focused on urban and neighborhood scale issues for different population groups. The relationship between building and site characteristics and participation in physical activity has not been explored in any detail.

This paper focuses on the role of environmental factors and their relationship with physical activity among older residents of CCRCs and other housing with services communities. In a separate paper (Harris-Kojetin, Kiefer, Zimring, Joseph, under review), we look at the role that social and organizational factors play in facilitating physical activity among retirement community residents.

Key Definitions

Physical activity has been defined as any *'bodily movement produced by the contraction of skeletal muscles that substantially increases energy expenditure, although the intensity and duration can vary'* (Singh, 2002, p. 263). It is important to make a distinction between 'physical activity' and 'activity.' While physical activity involves bodily movement and results in energy expenditure, an activity may or may not require bodily movement. Hence, reading, watching television, playing bingo are activities, though not physical activities. Walking, swimming, playing tennis or gardening are examples of physical activity.

The Surgeon General recommends at least 30 minutes of moderate intensity physical activity on most days of the week for health impact. The authors of the new recommendation on physical activity also suggest that physical activity benefits can be accrued in small bouts of regular household, occupational and leisure activities lasting at least 10 minutes at a time over the course of the day rather than necessarily in a single dedicated exercise session (Pate et al., 1995).

Research Questions

The findings presented here are part of a larger project that seeks to identify programs, practices and physical environmental features that promote physical activity in CCRCs and other senior housing with services settings. This project is a collaboration between the College of Architecture at the Georgia Institute of Technology and the Institute for the Future of Aging Services, an independent applied research center at the American Association of Homes and Services for the aging (AAHSA). This project was reviewed and approved by the Georgia Institute of Technology's Institutional Review Board.

The goals of this more specific inquiry are to:

1. Understand the extent of outdoor and indoor physical activity features and resources present in CCRCs and other senior housing providers to support physical activity among older adults
2. Identify how the presence and visibility of these physical activity features and resources may be related to physical activity participation levels among older adults in these communities.

The broad research question that emerges is: Is the presence and visibility of indoor and outdoor physical activity resources and features related to participation in physical activity among older adults in these communities? Throughout this paper, the independent living setting is abbreviated as IL, assisted living as AL, and nursing care as NC.

METHODS

Target Population

Continuing care retirement communities (CCRCs) are campus-type retirement communities offering a range of housing, services and health care that is centrally planned and administered. CCRCs are intended to supply a continuum of care (skilled nursing care, assisted living and independent living) throughout the lifetime of elderly residents. The majority of CCRCs offer all three levels of care. This allows residents to enter into the community while still relatively healthy and then move on to more intensive care as it becomes necessary (Sanders, 1997).

There are an estimated 2,600 CCRCs in the United States. There is no "universal" definition for CCRCs because individual states define what they are. Most CCRCs are located in urban or suburban locations–69% and 12%, respectively. About three-quarters are not-for-profit organizations (AAHSA, 2005). More than 660,000 Americans live in CCRCs. According to a 2004 survey of CCRCs by AAHSA, the average age of independent living CCRC residents is 83, compared to 87 for both assisted living and nursing care CCRC residents (AAHSA, 2005). Seventy-two percent of CCRC residents are female. Residents sign a contract with CCRCs articulating the specific housing and health services to be provided. These contracts come in several models, and range from moderate to expensive. The majority of CCRCs provide lifetime care in exchange for an upfront entrance fee and ongoing monthly fee.

Some provide an agreement that may be for a shorter period, however, with no upfront fee required.

Sample Frame Development and Sample Selection

The sample frame consists of not-for-profit providers in the membership of the AAHSA that provide more than one level of care–one of which is independent living (IL)–at the same address or at addresses within close geographic proximity. These providers are primarily CCRCs, but also include other IL housing providers offering at least one other level of care on the same campus. The final sample frame included 1,371 AAHSA CCRCs and housing providers meeting the above inclusion criteria. From the sample frame, we randomly selected 800 CCRCs and housing providers using SPSS statistical software.

Data Collection Design and Response Rate

Data collection occurred for eight weeks starting in January 2004. Surveys were sent via U.S. mail to prime contacts identified in the AAHSA membership database. Prime contacts were mainly Administrators, Assistant Administrators, CEOs and Executive Directors, and Directors of Nursing. We used a dual-mode approach, allowing respondents to complete the survey either by U.S. mail or web. Cover letters were sent with the mailed surveys and included a link to a web-based version of the survey questionnaire. The prime contacts were asked to direct specific survey questions to others in their facility as needed.

To ensure a favorable response rate and quality data, we implemented a multi-pronged data collection design–awareness messages about the upcoming survey using AAHSA's normal channels for communicating with members (e.g., web site, electronic memos, listservs), an advance letter, U.S. mailed questionnaire, reminder post card, e-mail reminders, phone call reminders, and the option to complete a web-based version of the survey.

A total of 463 surveys were returned (of the 800 in the random sample). Forty-one cases had to be excluded because their responses indicated that they did not meet inclusion criteria. Another 24 cases (all web survey completions) had to be excluded because of corrupt data (N = 10) or blank surveys being submitted (N = 14). In total, we had 398 valid respondents. The overall response rate is 52% (398/759).[1]

Description of Sample

The key characteristics of responding facilities are summarized in Table 1.

Survey Instrument Development

The survey instrument was developed using information collected through a literature review and informational interviews with CCRC management and staff and with architects that design retirement communities. We pretested the survey instrument with nine respondents from sites reflective of the target population, to gain insight into the substance of the survey (e.g., questions asked, definitions used) and the most effective ways to administer surveys and to increase response rates. The draft survey instrument was also sent to the project's Advisory Committee for comment. Input from pretest sites and committee members was compiled and used to refine the final draft of the survey instrument.

The final survey instrument contains 45 items, divided into four main sections to obtain the following information: (1) basic characteristics of responding campuses; (2) campus locations, grounds and outside community; (3) campus facilities and buildings; and, (4) campus residents and physical activity. Majority of the questions were close ended. Five open ended questions were included to obtain additional qualitative information (e.g., Please tell us some of the challenges your community has faced in getting your residents physically active). The paper survey is eight pages. The web version contained identical survey items. Only minor differences exist between the paper and web-based versions due to skip patterns and other web design issues. The survey takes about an hour to complete provided the information is at hand.

Research Design

In this paper we are focusing on describing available physical activity resources and their association with physical activity. The effect of personal factors (i.e., average age of residents) and organizational factors (i.e., number of organized physical activity programs offered on campus) on these relationships is also considered. The larger project also examined in detail the role of organized activities and of management structures and these findings are presented in another paper (Harris-Kojetin, Kiefer, Zimring, Joseph, under review).

TABLE 1. Key Characteristics of Responding Campuses

Characteristics of responding facilities		Distribution
Whether CCRC	CCRC	75%
	Non-CCRC	25%
Type of Contract[1]	Type A	25%
	Type B	22%
	Type C	43%
	No contract	6%
	Other	4%
Levels of care offered[2]	AL & IL	12%
	NC & IL	16%
	All three levels	72%
Average age of residents (years)	Independent Living residents	82
	Assisted Living residents	85
	Nursing Care residents	86
Average # of residents	Independent Living residents	157
	Assisted Living residents	45
	Nursing Care residents	82
Location[3]	Urban-large	14%
	Urban-small	27%
	Suburban	43%
	Rural	16%
Campus Size (acreage)	< 5 acres	16%
	5-25 acres	37%
	26-50 acres	24%
	51-100 acres	14%
	> 100 acres	9%
Campus age	1-10 years old	10%
	11-30 years old	39%
	31-40 years old	16%
	> 40 years old	35%
Campus terrain	Entirely flat	31%
	Mostly flat with some gradual slopes	51%
	Some hills	12%
	Very hilly	6%
Number of organized activities available on campus	Between 1-3 activities	37%
	Between 4-6 activities	33%
	Between 7-9 activities	21%
	Between 10-12 activities	9%
	Between 13-14 activities	< 1%

[1]Type A–extensive (lifetime); assisted living and skilled nursing costs included in basic fees; Type B–modified; some lifetime care benefits covered through basic fees, while other benefits offered at an additional charge, as needed; Type C–fee-for-service; all services offered on a pay-as-you-go basis, at a rate specified by the provider
[2] IL–Independent Living, AL–Assisted Living, NC–Nursing Care
[3] Urban-large–located within city limits of city with a population exceeding 500,000
Urban-small–located within city limits of a city with a population up to 500,000
Suburban–located within 50 miles of small or large urban population
Rural–no small or large urban population within 50 miles of the campus

Physical Environmental Variables

The paper focuses on the relationship between the presence and number of indoor and outdoor physical activity resources and features and participation in physical activity. Further, we examine whether visibility of outdoor resources present on campus is related to participation in physical activity. While the impact of visibility on physical activity has not been dealt with previously in any empirical studies within the physical activity research arena, case studies conducted in residential facilities for older adults suggest that this may be an important factor influencing an older adults decision to use outdoor spaces for physical activity (Regnier, 1994; Parker & Joseph, 2003). The independent variables considered in this study include:

1. Presence of specific outdoor features on campus including walking paths, swimming pools, golf course/putting greens, outdoor tennis court, resident garden plots, outdoor bowling areas, gardens, courtyards and porches with seating.
2. Number of outdoor features: This variable is the numerical sum of all outdoor features available on campus.
3. Visibility of outdoor features: For each of the outdoor features, respondents were asked: whether the feature was easily visible by many residents during daily activities (i.e., from apartments, public areas, or while walking on the campus).
   This information was provided by the key contact based on personal experience or input from other staff members. This may vary somewhat from resident experience but still provides a close estimate of features that are visible easily while walking around campus.
4. Presence of indoor physical activity facilities on campus including dedicated aerobics/exercise classroom, fitness room with equipment, indoor swimming pool, warm water therapy pool, indoor tennis courts, dance studio, indoor bowling alley, multipurpose activity room and dedicated physical therapy room.
5. Number of indoor physical activity facilities: This variable is the sum of all indoor physical activity facilities on campus

Outcome Variable

The key outcome described in this paper is resident participation in physical activity. The outcome measures described below were reported by respondents based on observed and recorded information available to them when completing the survey. While we have no independent confirmation of these numbers, the outcomes provide a measure of the general levels of physical activity among residents in a community and provide an idea of the degree of participation in specific activities such as swimming, golf, etc.

Respondents were asked to provide information on the average percentage of residents in each setting participating in different types of physical activities. We measure participation in physical activity in three ways:

1. Overall physical activity levels
2. Participation in particular physical activities
3. Walking to meals

Overall Physical Activity (PA) Levels: This outcome measures the percentage of residents (at three levels of care) that do at least 30 minutes of physical activity (PA) at least 3 times/week. This is based on physical activity guidelines that recommend at least 30 minutes of moderate intensity physical activity on most days of the week (Fletcher et al., 1996). The percentage of IL residents (43%) participating in PA for at least 30 minutes duration 3 times a week is almost twice the percentage of NC residents (23%) (Table 2). The literature suggests a decline in physical activity levels with functional ability and age, and that is borne out by these findings.

Activity PA: This is the average percentage of residents who participated in a particular physical activity at least once a week based on the question asking what percent of residents (at three levels of care) participated in a particular activity (13 items) at least once a week. Walking is by far the most popular activity among residents in all three settings, followed by aerobics and physical therapy (Table 2). As expected, for all physical activities there is a decline in participation levels from IL to AL settings. The only exception to this decline in participation is physical therapy, which is greatest among NC residents. This reflects the greater focus on providing restorative care to NC residents compared to IL and AL residents.

Walk to Meals: Since most communities offer a meal plan to residents as part of their monthly fee, walking to meals constitutes a regular instrumental activity. This outcome is based on a question that asks what percent residents (at three levels of care) walked to meals with or without assistance on a regular basis. Most IL and AL residents walk to meals on a regular basis. Only 29% of NC residents walk to meals regularly.

TABLE 2. Percentage of Residents in Different Settings Participating in Physical Activity

Outcome measure	IL residents (%)	AL residents (%)	NC residents (%)
Overall PA	43	32	23
Activity PA			
Walking on own	72	60	21
Walking as part of a club	7	4	2
Yoga/Pilates	2	1	0
Tai-chi/martial arts	3	1	1
Dance	4	1	1
Golf	5	1	0
Swimming (Indoor or Outdoor)	7	1	1
Shuffleboard	3	1	1
Bowling (Indoor or lawn)	3	2	3
Tennis (Indoor or Outdoor)	1	0	0
Aerobics	9	7	4
Water aerobics	5	1	0
Physical Therapy	7	9	20
Walk to meals	87	81	29

ANALYSIS

All data from the surveys were imported into SPSS statistical software version 13.1. The data were analyzed using different types of statistical techniques such as bivariate correlations, t-tests for significance of independent samples and linear regression.

Relationships between physical activity outcomes and variables of interest are reported in this paper only when the physical activity is prevalent in at least 5% of the responding communities. All relationships are statistically significant at 0.05 level or better. Statistically non-significant relationships are designated as NS in the tables presented in this paper.

RESULTS

Outdoor Features on Campus

More than two-thirds of the communities have paths, gardens, garden plots, courtyards and porches. Less than a third of the communities have outdoor swimming pools, golf courses, shuffleboard courts, bowling facilities and outdoor tennis (Table 3).

In most communities where the outdoor feature is present, it is easily visible during daily activities (e.g., from apartments, public areas, or while walking on campus). The only exception is tennis courts, which are easily visible in only 17% of the communities where they are available (Table 3). In terms of the number of outdoor features available on campus (Table 4), around a third of the campuses have 5 outdoor features, and 85% of the campuses have between 3 and 7 features.

Relationship Between Presence of Outdoor Features and Participation in Physical Activity

There appears to be a consistent association between the presence of an outdoor feature on campus and residents' participation in physical

TABLE 3. Presence and Visibility of Outdoor Features on Campus

Outdoor facility features	% of campuses where this feature is present	% of campuses where this feature is available and also visible
Swimming Pool	20	86
Paths	85	93
Golf Course	18	93
Outdoor Tennis	5	17
Gardens	77	89
Garden Plot	69	81
Shuffleboard Court	29	79
Bowling Area	11	89
Courtyard	83	90
Porch	82	89

TABLE 4. Number of Outdoor Features on Campus

Number of outdoor features on campus	% distribution of campuses
0	1
1	1
2	6
3	10
4	19
5	31
6	16
7	10
8	5
9	2

activity. Table 5 shows the participation levels in different activities among residents in campuses that have a certain outdoor feature and campuses that do not have the feature. Of specific interest is the fact that more IL residents participate in walking clubs on campuses that have walking paths (8% vs. 3%), gardens (14% vs. 6%) or outdoor lawn bowling areas (14% vs. 6%). Though the numbers are relatively small, the presence of walking paths is also related to more AL residents walking as part of a club (5% vs. >1%). These relationships are true even controlling for age of residents.

The presence of an outdoor swimming pool is related to more IL residents participating in swimming (17% vs. 4%) and water aerobics (11% vs. 3%). Interestingly enough, the one outdoor feature that appears to be related to many physical activity outcomes is the presence of a golf course. Eighteen percent of the campuses surveyed have golf courses and these communities clearly have more IL residents participating in many different activities. Also more residents (in all three levels of care) that live on campuses with golf courses are active for at least 30 minutes 3 times a week. These relationships remained significant even after controlling for differences in age of residents across campuses.

It should be noted that the number of features on a campus are highly correlated with the number of activity programs available on campus. Our initial informational interviews with campus administrators sug-

TABLE 5. Relationship Between Presence of Outdoor Feature and Participation in PA

Campuses where outdoor features are present more/less residents are likely to participate in.........	Indoor physical activity facility											
	Walking Paths		Outdoor Swimming Pool		Golf Course		Garden		Shuffleboard Court		Outdoor Lawn Bowling Area	
	No	Yes	No	Yes	No	Yes	No	Yes	No	Yes	No	Yes
Average % of IL residents engaging in...												
Walk on own	NS		NS		NS		NS		NS		NS	
Walking Club	3	8	NS		NS		6	14	NS		6	14
Aerobics	NS		8	15	8	15	4	11	8	12	8	17
Swimming	NS		4	17	6	12	5	8	6	10	NS	
Golf	NS		4	8	4	10	3	6	4	8	4	12
Dance	NS		NS		3	9	NS		3	7	3	11
Shuffleboard	NS		NS		NS		NS		1	8	NS	
Yoga	NS		NS		2	5	NS		NS		2	5
Bowling	NS		NS		NS		NS		NS		1	11
Water Aerobics	NS		3	11	4	8	3	5	NS		NS	
Physical Therapy	NS		6	10	8	10			NS		NS	
Average % of AL residents engaging in...												
Walking Club	>1	5	NS		NS		NS		NS		NS	
Aerobics	6	11	NS		6	11	NS		NS		NS	
Average % of IL residents participating in 30 minutes of PA 3 times a week	42	49	NS		41	53	NS		NS		NS	
Average % of AL residents participating in 30 minutes of PA 3 times a week	NS		NS		30	39	NS		NS		NS	
Average % of NC residents participating in 30 minutes of PA 3 times a week	NS		NS		22	30	NS		NS		NS	
Average % of IL residents walking to meals	NS		NS		85	92	80	88	NS		NS	
Average % of NC residents walking to meals	NS		NS		NS		22	31	NS		NS	

gested that such a recursive relationship may exist: more active communities are likely to create and maintain facilities while the presence of facilities allows physical activity programs to go on. When the number of physical activity programs offered on campus was introduced as a control variable many of the relationships became statistically non-significant. That is, campuses with golf courses are also likely to have many different types of program offerings which may influence resident participation in physical activity. This is consistent with an ecological model where many different factors together influence participation in physical activity.

Relationship Between Number of Outdoor Features on Campus and Participation in Physical Activity

Campuses with more outdoor features tend to have more residents at all three levels of care participating in different types of physical activities (Table 6), though the relationships are strongest for IL residents. Also, there is a significant correlation between number of outdoor facility features and

TABLE 6. Relationship Between Number of Outdoor Features on Campus and Resident Participation in PA

Campuses with more outdoor facility features on campus tend to have more residents participating in PA… (in descending order of strength of relationship among IL residents)	Pearson's R correlation		
	Independent Living (IL) residents	Assisted Living (AL) residents	Nursing Care (NC) residents
Average % of residents engaging in….			
Swimming	.34	.22	.12
Golf	.33	.14	NS
Tennis	.29	.15	NS
Water aerobics	.28	.17	.18
Aerobics	.28	.20	.13
Dance	.28	.13	NS
Yoga	.21	.16	NS
Bowling	.16	NS	NS
Walking clubs	.15	.12	NS
Tai chi	.14	NS	NS
Physical therapy	.11	NS	NS
Average % of IL residents participating in 30 minutes of PA 3 times a week	.16	NS	NS

participation in physical activity for as many as 11 activities for IL residents, 8 activities for AL residents and only 3 activities for NC residents. Also, more IL residents participate in at least 30 minutes of PA 3 times a week in campuses with more outdoor features. These relationships are significant even when controlling for age of residents. However, most of these relationships become statistically non-significant when the number of physical activity programs offered on campus is included as a control variable. The relationship between the number of outdoor facility features on campus and IL residents participating in golf, tennis and aerobics remains significant even when controlling for age and number of physical activity programs offered on campus.

Relationship Between Visibility of Outdoor Features and Participation in Physical Activity

We were interested in finding out if more people participated in specific activities when outdoor features related to that activity were present as well as visible during the course of daily activities. We found that where courtyards were visible, more IL residents walk as part of a club and where shuffleboard courts are visible more of IL residents participate in shuffleboard (Table 7). However, the actual level of participation in these physical activities is low in all communities.

Indoor Physical Activity Facilities

The communities surveyed reported having a range of different indoor physical activity facilities (Table 8). More than two-thirds of the

TABLE 7. Relationship Between Visibility of Specific Outdoor Facilities and Resident Participation in PA

Compared to campuses where outdoor facilities features are not visible, campuses that have visible outdoor facility features tend to have more residents engaging in PA....	Courtyard Visible?		Shuffleboard Court Visible?	
	Yes	No	Yes	No
Average % of IL residents engaging in...				
Walk as part of a club	7	1	NS	
Shuffleboard	NS		9	3

TABLE 8. Distribution by Type of Indoor Physical Activity Facilities Present on Campus

Indoor physical activity facility on campus	% distribution of campuses
Multipurpose activity room (used for many activities including, but not limited to, PA)	88
Fitness room with equipment	70
Dedicated physical therapy room/facility	67
Dedicated aerobics/exercise classroom	35
Indoor swimming pool	21
Warm-water therapy pool	18
Dance studio	4
Indoor bowling alley	2
None	2
Indoor tennis courts	< 1

communities have a multipurpose activity room (88%), fitness room with equipment (70%) and dedicated physical therapy room/facility (67%). Around 35% of the communities have dedicated an aerobics/ exercise room. Twenty-one percent of the communities have an indoor swimming pool and 18% have a warm-water therapy pool. Very few communities have indoor tennis courts, dance studios or indoor bowling alley.

Relationship Between Presence of Indoor PA Facility and Participation in PA

The presence of indoor physical activity facilities on campus was related to more residents participating in physical activity (Table 9). As expected, more residents participated in a particular activity if the related physical activity facility was present. For example, in campuses where an indoor exercise classroom or an indoor fitness room is present, resident participation in aerobics is almost double (at all three levels of care) that of campuses without these facilities. A similar trend is seen with the presence of indoor swimming pools or warm water therapy pools and participation in swimming and water aerobics; or presence of physical therapy room and participation in physical therapy. Though the presence of any particular indoor physical activity facility does not

Joseph et al. 159

TABLE 9. Relationship Between Presence of Indoor PA Facility and Participation in PA

Campuses where indoor PA facilities are present more/less residents are likely to participate in activities	Indoor physical activity facility											
	Indoor exercise classroom		Indoor fitness room		Indoor swimming pool		Indoor warm water therapy pool		Dance studio		Physical Therapy room	
	No	Yes	No	Yes	No	Yes	No	Yes	No	Yes	No	Yes
Average % of IL residents engaging in...												
Walk on own	75	66	NS		74	65	NS		NS		NS	
Aerobics	7	14	3	12	NS		NS		NS		NS	
Swimming	6	10	NS		4	16	6	14	7	16	5	8
Golf	4	7	2	6	5	7	NS		5	16	NS	
Dance	3	6	1	5	NS		NS		4	19	NS	
Shuffleboard	2	5	NS		NS		NS		NS		NS	
Yoga	NS		NS		NS		NS		2	7	NS	
Bowling	NS		NS		NS		NS		2	25	NS	
Tai chi	NS		NS		NS		NS		3	10	NS	
Water aerobics	4	7	2	6	3	11	4	10	NS		3	6
Physical Therapy	NS		4	7	6	9	NS		NS		4	8
Average % of AL residents engaging in...												
Aerobics	5	11	3	8	NS		NS		NS		NS	
Physical Therapy	NS		6	10	8	13	NS		NS		15	22
Average % of NC residents engaging in...												
Aerobics	2	6	NS		NS		NS		NS		NS	
Average % of IL residents walking to meals	NS		79	90	NS		NS		NS		NS	
Average % of NC residents walking to meals	NS		NS		NS		NS		NS		24	31

seem to be related to overall participation in PA (30 minutes of PA 3 times a week), the presence of indoor fitness rooms is associated with more IL residents walking to meals on a regular basis and the presence of physical therapy rooms with more NC residents walking to meals. This may be related to a stronger emphasis on fitness and restorative care in such communities.

Relationship Between Number of Indoor PA Facilities on Campus and Participation in PA

The number of indoor physical activity facilities present on campus is related to residents at all three levels of care participating in different types of PA (Table 10). Campuses with more indoor PA facilities tend to have more IL residents participating in swimming, water aerobics, golf, tennis, aerobics and physical therapy, dance, tai chi, bowling and yoga. However, fewer IL residents walked on their own in such campuses. Campuses with more indoor PA facilities also tend to have more AL residents participating in swimming, yoga and aerobics and NC residents participating in swimming, water aerobics, tennis, aerobics and

TABLE 10. Relationship Between Number of Indoor PA Facilities on Campus and Resident Participation in PA

Campuses with more indoor physical activity facilities on campus tend to have more residents participating in PA... (in descending order of strength of relationship among IL residents)	Pearson's R correlation		
	Independent Living (IL) residents	Assisted Living (AL) residents	Nursing Care (NC) residents
Average % of residents engaging in....			
Swimming	.47	.37	.22
Water aerobics	.41	NS	.12
Golf	.28	NS	NS
Tennis	.23	NS	.16
Aerobics	.20	.13	.12
Physical therapy	.19	NS	NS
Dance	.14	NS	NS
Tai chi	.13	NS	NS
Walk on their own	−.13	NS	NS
Bowling	.11	NS	NS
Yoga	.11	.16	NS
Walk as part of a club	NS	NS	NS
Shuffleboard	NS	NS	.14
Average % of residents walking to meals on a regular basis...	.12	NS	NS

shuffleboard. Especially of interest is the fact that the number of indoor PA facilities is strongly related to residents at all levels of care participating in swimming. The relationship between the number of indoor physical activity facilities on campus and participation in swimming (AL and IL) and water aerobics (IL and NC) exists even after controlling for the age of residents and number of physical activity programs offered on campus, though the relationship is weakened when these factors are taken into account.

DISCUSSION

Physical activity behavior is a complex phenomenon and several different factors influence participation in PA. This project is one of very few studies exploring the relation between building and site level factors and participation in physical activity among older adults in residential facilities. Our goal was to understand what types of resources were available in these campuses and whether the presence of resources could actually be linked to residents being physically active.

This article suggests that the presence of individual facilities and features as well as the actual number of facilities present is related to the resident participation in physical activity. Specifically, we found that campuses with certain outdoor features had more residents participating in different types of activities. For example, campuses with walking paths, gardens or outdoor lawn bowling areas had more independent living residents participating in walking clubs. We also found a significant relationship between the visibility of courtyards on campus and participation in walking clubs. This begins to suggest that in campuses where natural outdoor features are present on site and are visible, more people may participate in a social physical activity such as a walking club. The nature of the survey tool did not allow for further exploration of the connection between visibility and participation in physical activity. However the tentative findings from this study and observations from case studies that suggest that designing outdoor areas that are easily visible from public and circulation areas within buildings increases participation in physical activity deserve further exploration.

The study also found that campuses with more outdoor features are likely to have more residents participating in a range of different activities. This remained true after controlling for age of residents, but not when number of physical activity programs on campus was introduced as a control variable. A similar trend was seen with the presence and number of indoor physical activity facilities on campus and the participation in physical activities. The number of IL residents participating in swimming and water aerobics was significantly related to number of indoor physical activity facilities on campus even after controlling for age of residents and number of programs offered on campus, though the relationship was somewhat weakened. The study exemplifies the complexity of physical activity behavior and the difficulty in isolating the influence of the environment on physical activity behavior. CCRCs that encourage physical activity are likely to build and maintain more physical amenities that support physical activities as well as conducting a greater number of organized activities. As shown in this study, the relationship between physical design and PA was weakened or made insignificant when controlling for age or programming. This suggests that the interaction among person-level characteristics, physical design and organizational factors (programming) provides a better, if more complex, understanding of PA factors than looking at any one set of measures alone. The social ecological model posits just this—that these different factors influence physical activity behavior and also influence each other. As an exploratory study, this paper begins to identify how the availability of resources at the building and site level may be related to participation in physical activity. However, there is need for more focused studies that assess how the presence of resources at the building and site level influences participation in physical activity.

The nature of causation between the physical environment and activity is complex and we do not assert that if you build it they will come. Rather, the physical environment appears to be a facilitator that allows motivated staff and residents to work together to become more physically active.

LIMITATIONS

Several factors need to be considered in this study. While the AAHSA members of the team pre-tested the questionnaire for comprehension and relevance, this study depends on report of community managers and other staff which might not be accurate. Campus staff and management do not have a complete picture of the full range of activities in which residents participate. This is particularly true of IL residents, who are more likely to engage in physical activity off campus on their own. The study does not include the perspective of residents. The response rate is quite high for studies of this type, but remains only slightly over

50% and the nature of AAHSA's database did not allow us to compare the characteristics of responders and non-responders.

The physical activity outcome measure of at least 30 minutes a day for at least three times per week is a high criterion to set for this population. With this measure we do not capture physical activity levels of less frequency or duration. The list of specific physical activities used as outcome measures tends to focus more on programmed, organized physical activities. The list also excludes numerous other types of physical activities that older adults may participate in on a weekly basis (e.g., bicycling, gardening, etc.) that may be done alone or as part of a group.

FUTURE DIRECTIONS

Focus Groups with Residents

The survey reflects the perspectives of management and staff at communities. Focus groups with residents would provide insights about resident perceptions of the outdoor and indoor features available on campus. Further, focus groups with residents may help in better understanding how the visibility of outdoor and indoor features may encourage or motivate residents to be physically active.

In-Depth Case Studies

More in-depth case studies in a few communities would enable objective measurement of actual resident physical activity levels as well as resident use of physical activity resources and outdoor features. For example, participation in walking clubs is associated with the presence of outdoor landscaped areas on campus. The use of outdoor spaces for social physical activities such as walking clubs can be assessed in greater detail through case studies.

Checklist

The above activities and the survey results could inform the development of a comprehensive checklist for CCRCs and other housing providers of physical design features and programming that might encourage physical activity among residents. The checklist would provide a basis for providers and architects to assess their communities and to target potential changes that may increase physical activity.

NOTE

1. We computed the response rate according to the methods described in the American Association for Public Opinion Research's document, *Standard Definitions: Final Dispositions of Case Codes and Outcome Rates for Surveys*, 2004.

REFERENCES

AAHSA. (2005). The continuing care retirement communities 2004 profile. Washington, DC: Association of Homes and Services for the Aging (AAHSA).

Booth, M. L., Owen, N., Bauman, A., Clavisi, O., & Leslie, E. (2000). Social-cognitive and perceived environment influences associated with physical activity in older Australians. *Preventive Medicine, 31*(1), 15-22.

Brownson, R., Housemann, R. A., Brown, D. R., Jackson-Thompson, J., King, A. C., Malone, B. R. et al. (2000). Promoting physical activity in rural communities: Walking trail access, use, and effects. *American Journal of Preventive Medicine, 18*(3), 235-241.

Carnegie, M. A. B., Marshall, A.L. Mohsin, M. Westley-Wise, V. Booth, M.L. (2002). Perceptions of the physical environment, stage of change for physical activity, and walking among Australian adults. *Research Quarterly for Exercise & Sport, 73*(2), 146.

Fletcher, G., Balady, G., Blair, S., Blumenthal, J., Caspersen, C., Chaitman, B., Epstein, S., Sivarajan Froelicher, E., Froelicher, V., Pina, I., & Pollock, M. (1996). Statement on Exercise. Benefits and Recommendations for Physical Activity Programs for All Americans: A Statement for Health Professionals by the Committee on Exercise and Cardiac Rehabilitation of the Council on Clinical Cardiology. American Heart Association. *Circulation*, 94, 857-862.

Howell, S. (1980). Designing for aging: Patterns of use. Cambridge, MA: MIT Press.

King, A. C. (2001). Interventions to promote physical activity by older adults. *The Journals of Gerontology*, 56a (Supplement: Nutrition, Physical Activity, and Quality of Life in.), 34-46.

King, A. C., Rejeski, W. J., & Buchner, D. M. (1998). Physical activity interventions targeting older adults: A critical review and recommendations. *American Journal of Preventive Medicine, 15*(4), 316-333.

Parker, D., & Joseph, A. (2003). Creating environments to promote physical activity among older adults. Poster presented at the EDRA 34/2003 People Shaping Places Shaping People, Minneapolis.

Pate, R. & Pratt, M. (1995). Physical activity and public health. *JAMA: Journal of the American Medical Association* (Vol. 273, pp. 402): American Medical Association.

Regnier, V. (1994). Assisted living housing for the elderly: Design innovations from the United States and Europe. New York: Van Nostrand Reinhold.

Robert Wood Johnson Foundation (2000). The national blueprint on physical activity among adults age 50 and older.

Sanders, J. (1997). Continuing care retirement communities: A background and summary of current issues. Washington, DC: Office of the Assistant Secretary for Planning and Evaluation.

Satariano, W. A., & McAuley, E. (2003). Promoting physical activity among older adults: From ecology to the individual. *American Journal of Preventive Medicine, 25*(3, Supplement 2), 184-192.

Shephard, R. J. (1997). Aging, physical activity, and health. Champaign, IL: Human Kinetics.

Singh, M. A. F. (2002). Exercise comes of age: Rationale and recommendations for a geriatric exercise prescription. *The Journals of Gerontology, 57*A(5), M262-M282.

Takano, T., Nakamura, K., & Watanabe, M. (2002). Urban residential environments and senior citizens' longevity in megacity areas: The importance of walkable green spaces. *Journal of Epidemiology and Community Health, 56*, 913-918.

USDHHS. (1996). Physical activity and health: A report of the surgeon general. Atlanta, GA: U.S. Department of Health and Human Services, Centers for Disease Control and Prevention, National Center for Chronic Disease Prevention and Health Promotion.

Zimring, C., Joseph, A., Nicoll, G. L., & Tsepas, S. (2005). Influences of building design and site design on physical activity: Research and intervention opportunities. *American Journal of Preventive Medicine, 28*(2, Supplement 2), 186-193.

Environmental Support
for Outdoor Activities
and Older People's Quality of Life

Takemi Sugiyama
Catharine Ward Thompson

SUMMARY. The outdoor environment provides older people with opportunities to be active, have contact with nature and meet friends and neighbours. Research has shown that such outdoor activities have substantial benefits for older people's well-being. However, going outdoors is often difficult for them due to increasing frailty and environmental barriers. This study argues that a neighbourhood environment facilitating older people's outdoor activities has a positive effect on their well-being. Small-scale studies were conducted to explore the concept

Takemi Sugiyama, PhD, is Research Fellow, OPENspace Research Centre, Edinburgh College of Art, 79 Grassmarket, Edinburgh EH1 2HJ, UK (E-mail: t.sugiyama@eca.ac.uk).

Catharine Ward Thompson, FLI, FRSA, is Professor and Director, OPENspace Research Centre, Edinburgh College of Art, 79 Grassmarket, Edinburgh EH1 2HJ, UK (E-mail: c.ward-thompson@eca.ac.uk).

The authors are very grateful for helpful comments from the anonymous reviewers and Professor Peter Aspinall in finalising the paper.

This study is part of a research project called I'DGO (Inclusive Design for Getting Outdoors), which is supported by the Engineering and Physical Sciences Research Council. The I'DGO project is being undertaken by a partnership of Edinburgh College of Art and Heriot-Watt University, Oxford Brookes University and the University of Salford. The Sensory Trust, RICAbility, the Housing Corporation and Dementia Voice are non-academic partners in the I'DGO consortium.

[Haworth co-indexing entry note]: "Environmental Support for Outdoor Activities and Older People's Quality of Life." Sugiyama, Takemi, and Catharine Ward Thompson. Co-published simultaneously in *Journal of Housing for the Elderly* (The Haworth Press, Inc.) Vol. 19, No. 3/4, 2005, pp. 167-185; and: *The Role of the Outdoors in Residential Environments for Aging* (ed: Susan Rodiek, and Benyamin Schwarz) The Haworth Press, Inc., 2005, pp. 167-185. Single or multiple copies of this article are available for a fee from The Haworth Document Delivery Service [1-800-HAWORTH, 9:00 a.m. - 5:00 p.m. (EST). E-mail address: docdelivery@haworthpress.com].

Available online at http://www.haworthpress.com/web/JHE
© 2005 by The Haworth Press, Inc. All rights reserved.
doi:10.1300/J081v19n03_09

of "environmental support" for outdoor activities and its effect on people's Quality of Life (QOL). Two methods were employed to identify the degree of environmental support. One was based on the assessment of neighbourhood environments and the other focused on outdoor activities people undertake. Analysis found highly significant correlations between environmental support and participants' QOL in both measurements even after controlling for participants' functional capability. The results suggest that outdoor environments adjacent to where one resides play a significant role in older people's QOL. *[Article copies available for a fee from The Haworth Document Delivery Service: 1-800-HAWORTH. E-mail address: <docdelivery@haworthpress.com> Website: <http://www.HaworthPress. com> © 2005 by The Haworth Press, Inc. All rights reserved.]*

KEYWORDS. Older people, outdoor environment, outdoor activity, quality of life, neighbourhood

INTRODUCTION

The outdoor environment offers great opportunities for older people to be physically active, to have contact with nature and to meet with friends and neighbours. However, it also presents various barriers that prevent them from going out. Due to the combination of increasing frailty in late life and barriers in the environment, going outdoors is often the first set of activities that older people find too hard to perform (Shumway-Cook et al., 2003). The sedentary life style that often results is considered a serious health risk for older people (WHO, 2003). Thus, it can be argued that an outdoor environment that makes going out easy for older people plays an important role in maintaining and enhancing Quality of Life (QOL) in late life. Environmental factors in people's participation in physical activity have started to capture the attention of researchers in public health (e.g., Booth et al., 2000; Giles-Corti & Donovan, 2002; Humpel et al., 2004; Li et al., 2005; Saelens, Sallis, Black & Chen, 2003; Satariano & McAuley, 2003). Older people's "mobility" in the outdoor environment and its implications on their well-being have also been discussed in some studies (e.g., Metz, 2000; Mollenkopf et al., 2004). However, little research effort has been directed at understanding the role of the outdoor environment in older people's QOL (Wahl & Weisman, 2003).

As an initial attempt to explore the concept of environmental support, the present study briefly reviews the benefits of outdoor environment on older people's well-being. Then it proposes environmental support for outdoor activities as a key concept to understand the effects of the outdoor environment on older people's activity and well-being. After discussing its theoretical background and measurement methods, the paper presents the results of small scale pilot studies that examined the association between environmental support and older people's QOL.

BENEFITS OF OUTDOOR ENVIRONMENTS FOR OLDER PEOPLE

The literature suggests that the benefits of outdoor environments on older people are obtained from three different types of engagement with outdoor environments. They are (1) participation in outdoor physical activity, (2) exposure to outdoor natural elements and (3) social interaction with friends and neighbours in outdoor places (Bowling et al., 2003; de Vries, Verhaji, Groenewegen & Spreeuwenberg, 2003; Nezlek, Richardson, Green, & Schatten-Jones, 2002). The following section briefly reviews in what way outdoor environments contribute to QOL in late life.

Benefits from physical activity. Abundant evidence indicates that regular participation in moderate physical activities has substantial benefits for the health of older people. A physically active lifestyle is found to minimise the physiological changes associated with ageing and help delay or prevent the onset of common chronic diseases (Singh, 2002). Keysor and Jette (2001) have also shown in their review that participation in physical activity improves older people's physical condition, including muscle strength, aerobic capacity, balance and flexibility. Such enhancement is known to help reduce the possibility of falling, which is a major cause of disabilities for older people (Skelton, 2001). Research has also demonstrated that participation in physical activity has protective effects against insomnia (e.g., Morgan, 2003). Studies addressing this topic do not normally differentiate indoor and outdoor activity. However, since walking is considered one of the most common and accessible activities (e.g., Department of Health, 2004), it can be argued that the benefits discussed here are applicable to outdoor activities.

In addition to the health benefits, regular physical activity provides older people with psychological benefits. Silverstein and Parker (2002) found that older Swedes who increased activity participation in a 9-year period showed an increase in life satisfaction. Another line of research examines the effect of physical activity on depression. A prospective study has identified that physical activity such as a long walk can reduce the risk of subsequent depression (Strawbridge, Delger, Roberts & Kaplan, 2002). The benefits of physical activity on cognitive functioning have also been demonstrated. Weuve et al. (2004) have shown that a higher level of physical activity (walking for more than 1.5 hours per week) is associated with better cognitive performance and memory in older women.

Benefits from contact with nature. Extensive research has shown the restorative effects of the natural environment (Kaplan, 1995). A classic study by Ulrich et al. (1991) showed that exposure to a 10-minute video of natural settings (after viewing a stressful film) brought faster and more complete stress recovery in comparison to the same length video of urban settings. Recent research has also found that the amount of time people spend in open green spaces is associated with a reduced risk of developing stress-related illnesses (Grahn & Stigdotter, 2003). Similarly, Hartig et al. (2003) indicated that those who walked in a natural setting exhibited increase in positive affect and decreased anger compared with those who walked in a built-up urban environment.

Several studies have explored the effects of neighbourhood green spaces on health. A longitudinal study in Japan investigated the association between older people's longevity and the existence of green areas that are nearby and easy to walk to (Takano, Nakamura & Watanabe, 2002). The authors found that the five-year survival percentage of older people who live in an area with such green spaces is significantly higher than those living in an area without such spaces. Another study in the Netherlands has shown that the amount of green in a neighbourhood is positively associated with health status of older people measured as the number of recent illnesses (de Vries et al., 2003). The authors have reported that the effect of green spaces on health is stronger for older people, whose outdoor exposure is more likely to be limited to neighbourhood environments.

Benefits from social interaction. Outdoor open spaces can serve as a place for social interaction among neighbours. It has been shown that the "greenness" of an open space invites more frequent use of the space by neighbours, and thus fosters stronger social ties among them (Kuo, Sullivan, Coley & Brunson, 1998). A study in Ireland has also found that people living in mixed-use, pedestrian-oriented neighbourhoods, which offer a greater chance to meet each other outdoors, tend to engage in social activities more often than those living in car-dependent neighbourhoods (Leyden, 2003). Since the planning and design of neighbourhood environments influence the way people interact informally in outdoor spaces, environmental factors have a considerable impact on the quantity/quality of informal social contacts among neighbours (Kuo et al., 1998).

The benefits of social interaction on older people are well documented. Bennett (2002) showed that a low level of social engagement in late life is correlated with a decline in physical health and high risk of mortality. Diversity in social relations and frequent social participation have been found to serve as protection against the onset of mobility disability (Avlund, Lund, Holstein & Due, 2004). Furthermore, a longitudinal study has found that frequent participation (daily-weekly) in social activities is associated with a decreased risk of dementia (Wang, Karp, Winblad & Fratiglioni, 2002). These studies did not specifically address the social interaction among neighbours. However, this type of locally-based social interaction can be particularly important for older people, because they are likely to spend more time around their house. In fact, a UK study found that older people consider having good relationships with their neighbours as an important constituent of their QOL (Bowling et al., 2003).

The definition of outdoor activity. The above overview illustrates that activities taking place in outdoor environments can enhance older people's well-being in several ways. The activity discussed here is not necessarily physically vigorous. To enjoy the benefits from contact with nature and social interaction, one only needs to go out and stay outside for a while. In relation to this point, McAuley et al. (2000) have demonstrated that both aerobic and nonaerobic activities have positive effects on older people's psychological well-being. Thus, the current study defines outdoor activity simply as "being outdoors," to encompass all the types of engagement with outdoor environments. With regard to the environment, a focus is placed on a neighbourhood environment, which is likely to be the most immediately accessible outdoor environment for older people. It is known that a poor quality environment may be one of the factors that deter older people from being active (e.g., Humpel, Owen & Leslie, 2002; Schutzer & Graves, 2004; Trost et al., 2002). Thus it can be argued that a neighbourhood environment that makes being outdoors easy and enjoyable is likely to encourage more outdoor activities, which in turn is conducive to a better QOL.

ENVIRONMENTAL SUPPORT FOR OUTDOOR ACTIVITY

The support function of the environment is obviously important for older people to remain active and independent. According to Lawton

(1986), this is one of the three vital functions the environment has to offer the older population: maintenance, stimulation and support. Past studies have explored the support function of the environment based on the idea of person-environment (P-E) fit, which refers to the congruence between the demand of the environment ("environmental press") and people's competence (Lawton & Nahemow, 1973). Kahana et al. (2003), for instance, discussed the significance of the P-E fit in the context of older people's residential satisfaction. Iwarsson (2005) has found that the P-E incongruence in housing is associated with residents' ADL (activities of daily living) dependence. However, environmental support for outdoor activities and its effect on older people's QOL remain to be explored (Wahl & Weisman, 2003).

One way of assessing environmental support draws on the appraisal of environmental characteristics. Research has documented scores of environmental factors that influence people's participation in activities (e.g., Booth et al., 2000; Giles-Corti & Donovan, 2002; Humpel et al., 2004). A number of instruments to assess supportiveness of the environment have been developed and reported based on the findings from existing empirical studies. This type of instrument can employ either objective measures (e.g., Pikora et al., 2002) or self-reported and subjective measures (e.g., Saelens et al., 2003). An advantage of this type of measurement is its direct connection with the environment. Findings obtained from studies employing such an instrument may be directly translated into design guidelines and policy recommendations. The measurements in this category may also work well in a neighbourhood level analysis. A drawback of this measurement method is that salient environmental attributes are assumed constant across people. Environmental attributes that influence the pattern of outdoor activities may differ between people who have different lifestyles and different functional capabilities. Some items in such scales may have little relevance to particular individuals or groups of people.

An alternative way of measuring environmental support uses people's activity as a unit of analysis. The degree of environmental support depends not only on environmental factors but also on personal factors such as the type of activity engaged in and a person's functional capability. It seems advantageous to focus on an individual's activity in addressing environmental support, because (1) older people are likely to be diverse in their choice of outdoor activities, and (2) activity carries both personal and environmental dimensions. The idea of making use of activities as an interactional unit of analysis originates from the concept of "personal projects" developed by Little (1983). Personal projects are a set of goal-oriented, self-generated activities that reflect an individual's construal of self and the context within which activities take place (Little, 2000). They are relevant to an individual's well-being, for project pursuit can be deemed as a process in which a person strives for his or her personal goals (Little & Chambers, 2004). Being engaged in meaningful projects and approaching personal goals through such projects are an important aspect in one's life (Omodei & Wearing, 1990). Since older people's activities are more likely to be subject to contextual constraints, this concept seems adequate to explore the relationship between the environment and older people.

RESEARCH AIM AND METHOD

Small-scale pilot studies were carried out to develop instruments to measure environmental support and to identify to what extent environmental support accounts for an individual's QOL. The study also examined the effects of an individual's functional capability on the relationship between the two constructs. It was anticipated that people's functional status may be associated with the way they perceive their surroundings and their well-being. Thus, it may confound the relationship between environmental support and QOL. In addition, since 'younger old' and 'older old' people are known to have different psychological profiles (e.g., Smith & Baltes, 1997), the study explored whether the effects of environmental support on QOL vary between different age groups. It was postulated that environmental support has a larger bearing on QOL for the older group, because a decrease in competence as a result of ageing calls for more supportive environments in order that daily activities may be performed.

Fifty-eight people aged 65 and older were recruited through colleagues and acquaintances in Edinburgh, Glasgow, Stockport (a suburb of Manchester) and Cornwall. Table 1 shows the age and gender distribution of the participants. They were asked to complete a questionnaire, which included questions on environmental support, QOL, functional capability and sociodemographic data. Since the instruments were developed in an incremental and iterative process, the questionnaire was slightly different at different stages. The later version of the questionnaire, for example, included questions on participants' general health status and outdoor activity pattern. The results shown below only deal with the common elements among different versions of the questionnaire.

TABLE 1. Age and Gender Distribution of the Participants

	Female	Male	Total
65-74	14	10	24
75-84	20	5	25
85+	3	0	3
Unknown	1	5	6
Total	38	20	58

Environmental support was measured in two ways. One measure was perceptual evaluation of neighbourhood environments. An 18-item scale was developed based on the focus group interviews conducted earlier, the instruments produced by Saelens et al. (2003) and Humpel et al. (2004), and various relevant design guidelines (e.g., Civic Trust, 2004; DTLR, 2002). Of 18 items in this scale, 3 items are relevant to outdoor spaces around one's house (e.g., "There is a pleasant place to sit outside the home where I live."), 11 of them are concerned with a local open space such as a park and routes to reach such a space (e.g., "The local open space is clean and well maintained." "The paths to get to the local open space are easy to walk on."), and 4 items ask about the larger neighbourhood area (e.g., "Steep hills and steps in my neighbourhood make it difficult to get around."). The scale focuses on natural or green environments because of the distinctive benefits (restorative and social) they possess for older people. The reliability (internal consistency) of the entire scale was 0.73.

The other way of identifying environmental support was based on personal projects. The original version of the personal projects questionnaire, which can be complex and lengthy for older people, was simplified for this study. The participants were asked to list outdoor activities they do regularly, have decided to undertake, or are thinking about doing (free description). Some examples such as "make my garden beautiful," "walk the dog everyday," and "play bowls" were given to suggest that it is "everyday" activity that is in question. They were then asked to evaluate each activity in terms of the extent to which the environment makes it difficult/easy to carry out, and its personal importance on a 5-point scale. In addition to the listed volitional activities, they were asked to rate "just go for a walk" on the same basis. Overall environmental support for a participant was calculated as a weighted means of support (difficulty/easiness) for the listed activities using the importance as a weight (Wallenius, 1999).

The outcome variable of the study was participants' QOL. Their life satisfaction was used as an indicator of QOL in this study. A 5-item Satisfaction With Life Scale (SWLS) developed by Diener and his colleagues (1985) was employed for this purpose. The reliability of the scale was 0.87. To assess the functional capability of participants, they were asked to indicate the ease with which they could perform six instrumental activities of daily living (IADLs) (Jette et al., 1986). The IADLs employed were mostly concerned with mobility such as walking a certain distance, climbing stairs and using public transportation. In addition, the number of outdoor activities (personal projects) the participants listed was also included in the analysis.

RESULTS

Table 2 shows the mean and standard deviation of the five variables for two age groups: people aged 65-74 and people over 75. LS (life satisfaction) is the mean of the responses to the five items in SWLS. The variable ranges from 1 (least satisfied) to 5 (most satisfied). ESP and ESN are environmental support based on personal projects and neighbourhood environments respectively. In both variables, the score ranges from 1 (least supportive) to 5 (most supportive). FC (functional capability) is the average degree of ease in performing the IADLs, which also ranges from 1 (lowest in functional capability) to 5 (highest in functional capability). NOA is the number of outdoor activities the participants listed voluntarily in response to the questions on ESP. The t-tests

TABLE 2. Means and Standard Deviations of the Variables by Age Group

	65-74	75+	T test
Life Satisfaction (LS)	3.84 (0.69)	3.89 (0.97)	*ns*
Environmental Support Projects (ESP)	4.28 (0.73)	3.71 (1.10)	$p < .05$
Environmental Support N'hood (ESN)	3.70 (0.76)	3.67 (0.94)	*ns*
Functional Capability (FC)	4.52 (0.85)	3.51 (1.37)	$p < .01$
Number of Outdoor Activities (NOA)	4.32 (1.49)	4.04 (1.15)	*ns*

(Note: standard deviations are shown in brackets)

showed that LS, ESN and NOA did not differ significantly between the age groups, but ESP and FC were significantly different between them. No significant differences were found in these variables according to sex of participants.

Table 3 shows bivariate correlation coefficients between the five variables. As can be seen, all the variables were highly correlated with each other except for two coefficients involving NOA. The analysis found high correlation coefficients between LS and ESP ($r = .55$, $p < .001$) and ESN ($r = .57$, $p < .001$). This means that environmental support accounts for about 30% of the total variance in life satisfaction. A significant correlation was found between ESP and ESN ($r = .48$, $p < .001$). This signifies that they share a certain amount of common variance, suggesting that they measure different facets of the same overall construct. Table 3 also shows that participant's FC was strongly correlated with ESP ($r = .63$, $p < .001$). This means that those who have difficulty in performing the IADLs tend to perceive their surroundings less supportive. The study posited that ESP is an interactional variable embracing both environmental and individual dimensions. The results shown here corroborated the interactional nature of ESP. In the case of ESN, however, the involvement of functional capability was smaller ($r = .39$, $p < .01$). ESN measures supportiveness focusing on neighbourhood natural environments, which is less dependent on individual level factors.

The number of activities participants listed (NOA), which indicates the diversity of activities a person conducts outdoors, can be envisaged as a surrogate of activeness of the person. NOA was found to be correlated with life satisfaction ($r = .43$, $p < .01$) and with ESN ($r = .48$, $p < .001$). This indicates that respondents who have a wider range of out-

TABLE 3. Bivariate Correlation Between the Variables

	LS	ESP	ESN	FC	NOA
Life Satisfaction (LS)	1				
Environmental Support Projects (ESP)	.55***	1			
Environmental Support N'hood (ESN)	.57***	.48***	1		
Functional Capability (FC)	.40**	.63***	.39**	1	
Number of Outdoor Activities (NOA)	.43**	.26	.48***	.22	1

*$p < .05$, **$p < .01$, ***$p < .001$

door activities tend to be more satisfied with life, and live in more supportive neighbourhood environments. However, the correlation between NOA and ESP was not significant ($r = .26$). A later section discusses possible reasons for this result.

Functional capability (FC) was correlated significantly with LS, ESP and ESN as shown in Table 3. This means that participants who have a better functional capability are likely to perceive their surroundings more supportive, and also likely to be more satisfied with life. Thus it is possible that the relationship between life satisfaction and environmental support is confounded by FC. To examine whether the relationship is spurious, partial correlation between LS, ESP and ESN controlling for FC was examined. Table 4 shows the results. Although the partial correlation coefficients were slightly smaller than the bivariate correlation, the significant correlation in the bivariate analysis remained significant in the corresponding partial correlation. The findings demonstrate that the relationship of environmental support with life satisfaction held even when the effects of participants' functional capability were removed.

As Table 2 indicates, the younger and older groups of participants differed in their functional capability. Difficulties in mobility and functioning as a result of ageing may alter the extent to which environmental support and well-being are associated. To explore how age modifies the relationship between the two, bivariate correlation was calculated separately for the two age groups. Tables 5 and 6 show the correlation coefficients for these groups.

Table 5 and 6 illustrate that the strong association between ESP and LS was constant across the younger ($r = .62, p < .01$) and older age groups ($r = .61, p < .01$). A slightly lower correlation was found between ESN and LS for people over 75 ($r = .53, p < .01$) in comparison to

TABLE 4. Partial Correlation Controlling for FC

	LS	ESP	ESN	NOA
Life Satisfaction (LS)	1			
Environmental Support Projects (ESP)	.44**	1		
Environmental Support N'hood (ESN)	.50***	.35*	1	
Number of Outdoor Activities (NOA)	.40**	.17	.43**	1

*$p < .05$, **$p < .01$, ***$p < .001$

TABLE 5. Bivariate Correlation for People Aged 65-74 (n = 24)

	LS	ESP	ESN	FC	NOA
Life Satisfaction (LS)	1				
Environmental Support Projects (ESP)	.62**	1			
Environmental Support N'hood (ESN)	.64**	.26	1		
Functional Capability (FC)	.13	.30	.43*	1	
Number of Outdoor Activities (NOA)	.43*	.45*	.49*	.46*	1

*$p < .05$, **$p < .01$, ***$p < .001$

TABLE 6. Bivariate Correlation for People Over 75 (n = 28)

	LS	ESP	ESN	FC	NOA
Life Satisfaction (LS)	1				
Environmental Support Projects (ESP)	.61**	1			
Environmental Support N'hood (ESN)	.53**	.54**	1		
Functional Capability (FC)	.51**	.68***	.37	1	
Number of Outdoor Activities (NOA)	.46*	.12	.45*	.16	1

*$p < .05$, **$p < .01$, ***$p < .001$

that in the younger group ($r = .64$, $p < .01$). Unlike the initial expectation, the finding suggests that neighbourhood natural environments are less relevant to the life satisfaction of people in the older group. The difference between the two groups was also found in the correlation between ESP and ESN. The older group showed a relatively large correlation coefficient ($r = .54$, $p < .01$), whereas it was non-significant in the younger group ($r = .26$). The difference of one's area of activity between the age groups may account for the difference. Namely, outdoor activities (a basis for ESP) listed by the younger group may predominantly take place in a wider area than immediate neighbourhood environments. The age groups also differed in the correlation involving functional capability. For the older group, a significant correlation was found between LS and FC ($r = .51$, $p < .01$), while the same correlation for the younger group was non-significant ($r = .13$). The correlation between FC and ESP was also different between the two groups: highly significant for the older group ($r = .68$, $p < .001$) and non-significant for the younger group ($r = .30$). These results can be understood as the growing importance of functional capability in life satisfaction and in perceptions of the supportiveness of the environment as one becomes very old.

DISCUSSION

The main objective of the pilot studies was to examine whether the concept of environmental support is relevant to older people's QOL. Although the small sample size makes it difficult to draw firm conclusions, the results show that environmental support for outdoor activities explains about 30% of the variance in participants' life satisfaction. This indicates that the supportiveness of the outdoor environment plays a highly important role in older people's well-being. The results are consistent with a previous study conducted in Finland, which employed middle-age people as participants and the general environment (indoor and outdoor) as its scope of study (Wallenius, 1999). However, in comparison to the current study, environmental support accounted for much smaller variance in participant's life satisfaction in the Finnish study. It can be argued that the stronger association between environmental support and life satisfaction was obtained in this study, because older people are more vulnerable to environmental barriers, especially in outdoor environments.

The findings of the study provided evidence in favour of environmental support as a concept linking the environment and people's QOL. Firstly, environment support measured through two different instruments, which have different theoretical origins, was similarly correlated with life satisfaction. The fact that two different approaches produced significant association suggests a level of robustness in this conceptualisation. Secondly, partial correlation analysis excluded the likelihood that the relationship between environmental support and well-being was confounded by a participant's functional capability. The analysis eliminated a major plausible cause of spurious correlation. Thirdly, the strong correlation between the two variables was observed in both the younger and older age groups. Despite some age-related differences found in the analysis, environmental support was consistantly correlated with participants' life satisfaction in both age groups. Fourthly, the findings indicated significant correlations between environmental support (ESN) and the number of outdoor activities. Since more outdoor activities mean more benefits, the association between them is likely to reinforce the link between environmental support and well-being. Lastly, environmental support was more closely associated with life satisfaction compared to functional status, particularly in the younger old people, which further substantiates the significance of the environmental dimension in their QOL. These findings appear to demonstrate that the concept of environmental support is highly effective in capturing aspects of the outdoor environment that are relevant to older people's QOL.

The comparison of the two age groups generated interesting findings. As shown above, the younger group showed a higher correlation between ESN and LS than the older group. Initially, it was expected that environmental support would matter more to people in the older group, who are more vulnerable to barriers imposed by the environment. Less frequent use of neighbourhood natural environments by the older participants might be a reason for this, but people's attitudes towards outdoor activity may also play a part in the results. Explaining the "environmental proactivity hypothesis," Lawton (1989) showed that the greater the functional competence of the person, the more likely the person actively seeks environmental resources that enable him or her to meet personal needs and wants. It can be argued that the younger participants may be more proactive and accordingly place a higher value on the quality of neighbourhood environments. Another notable difference between the groups is the relationship between ESP and NOA. The correlation was significant for the younger group ($r = .45, p < .05$), but not significant for the older group ($r = .12$). The small correlation in the older group can be considered as a reason for the non-significant correlation in the data overall between ESP and NOA. It is possible to assume that the participants in the older group have a certain number of basic activities that have to be done regardless of the degree of environmental support for these activities. (As shown in Table 2, the variance of NOA of the older group is smaller than that of the younger group.) On the other hand, the younger participants may be more varied with regard to the choice of outdoor activities, in which the supportiveness of the environment plays a relatively larger role.

A few theoretical and methodological issues in this study deserve further discussion. In this article, environmental support was considered to be a correlate (rather than cause) of an individual's QOL. There are two arguments against a causal relationship. First, older people who have a high level of well-being may be healthy and mobile and thus likely to perceive outdoor environments as easier to move around in. In this case, so long as the environment remains constant, it is QOL that influences environmental support. Second, it is probable that people whose QOL is higher (healthier, more active and possibly more affluent) choose to live in a more activity-friendly neighbourhood, which may involve a move later in life especially after retirement. Methodological limitations of this study include the small size of the sample, which is mostly urban, reliance on self-report data in measuring environmental support and other variables, and the omission of potentially important variables such as health status and the pattern of outdoor activity, which may be related both to environmental support and well-being. A large-scale study addressing these points is currently under way to substantiate the findings of this study.

Future studies may explore the concept of environmental support further to gain a richer understanding as to how outdoor environments are involved in older people's quality of life. Our review identified three types of engagement with outdoor environments that confer various benefits on older people: participation in physical activity, exposure to natural elements and social interaction with friends and neighbours. Each seems to have a unique contribution to make to a person's well-being. It would be worthwhile to study which type of engagement is more influential in the well-being of older people. Findings from such research would offer practical insights to inform policy making and planning of the outdoor environment. Another important future research focus is the identification of specific environmental attributes that are relevant to environmental support. This process requires identifying patterns of activities taking place in a setting and environmental attributes in the setting that affect those activities, then finding out which attributes have higher leverage in facilitating or hindering the activities. Information obtained from such investigation is obviously relevant to the design and management of environmental interventions that aim to encourage and enhance older people's outdoor activity. The role of proactive attitudes in the relationship between environmental support, QOL, and functional capabilities also merits further research. The findings imply that participants' proactive attitudes may have a bearing on the salience of environmental support. Potential research topics in this regard include whether supportive neighbourhood environments encourage more proactive attitudes among residents. Finally, it would be useful to investigate how environmental support for outdoor activities can contribute to the idea of "ageing-in-place." This concept normally refers to the home environment and is often discussed in association with interior spaces, e.g., home modification or assisted living (e.g., Ball et al., 2004). However, the outdoor environment is an important component for older people to remain independent. The instruments developed here offer valuable ways to explore the role of outdoor environments in ageing-in-place, and thus may contribute to the development of a better environmental policy for older people.

CONCLUSIONS

The concept of environmental support was proposed in the present study in order to make it possible to examine the effects of the environment on people's well-being. The literature review suggested that being outdoors can confer psychological and physiological benefits on older people. Thus, it was postulated that the environment that facilitates being outdoors would enhance QOL in late life. The results obtained in the pilot studies, i.e., the highly significant correlations between environmental support and participants' life satisfaction, can be interpreted as evidence that sustains this hypothesis. It can be inferred from this exploratory study that outdoor environments adjacent to where one resides play a significant role in one's quality of life. The findings of this research As far as the authors are aware, is the first attempt to assess the direct influence of the outdoor environment on an individual's QOL. The results from this research indicate a need for larger and more comprehensive studies to investigate the significance of the outdoor environment for the older population.

REFERENCES

Avlund, K., Lund, R., Holstein, B. E., & Due, P. (2004). Social relations as determinant of onset of disability in aging. *Archives of Gerontology and Geriatrics, 38*(1), 85-99.

Ball, M. M., Perkins, M. M., Whittington, F. J., Connell, B. R. et al. (2004). Managing decline in assisted living: The key to aging in place. *Journal of Gerontology: Social Sciences, 59B*(4), S202-S212.

Bennett, K. M. (2002). Low level social engagement as a precursor of mortality among people in later life. *Age and Ageing, 31*(3), 165-168.

Booth, M. L., Owen, N., Bauman, A., Clavisi, O., & Leslie, E. (2000). Social-cognitive and perceived environment influences associated with physical activity in older Australians. *Preventive Medicine, 31*(1), 15-22.

Bowling, A., Gabriel, Z., Dykes, J., Dowding, L. M., Evans, O. et al. (2003). Let's ask them: A national survey of definitions of Quality of Life and its enhancement among people aged 65 and over. *International Journal of Aging and Human Development, 56*(4), 269-306.

Civic Trust (2004). *Green flag award guidance manual*. Retrieved February 2005, from http://www.greenflagaward.org.uk/manual/

Department for Transport Local Government and the Regions (2002). *Green spaces, better places*. Retrieved February 2005, from http://www.odpm.gov.uk/stellent/groups/odpm_urbanpolicy/documents/page/odpm_urbpol_607952.pdf

Department of Health (2004). *At least five a week: Evidence on the impact of physical activity and its relationship to health*. London: Department of Health.

De Vries, S., Verheij, R. A., Groenewegen, P. P., & Spreeuwenberg, P. (2003). Natural environments–healthy environments? An exploratory analysis of the relationship between greenspace and health. *Environment and Planning A, 35*(10), 1717-1731.

Diener, E., Emmons, R. A., Larsen, R. J., & Griffin, S. (1985). The satisfaction with life scale. *Journal of Personality Assessment, 49*(1), 71-75.

Giles-Corti, B., & Donovan, R. J. (2002). The relative influence of individual, social and physical environment determinants of physical activity. *Social Science and Medicine, 54*(12), 1793-1812.

Grahn, P., & Stigdotter, U. A. (2003). Landscape planning and stress. *Urban Forestry and Urban Greening, 2*(1), 1-18.

Hartig, T., Evans, G. W., Jamner, L. D., Davis, D. S., & Garling, T. (2003). Tracking restoration in natural and urban field settings. *Journal of Environmental Psychology, 23*(2), 109-123.

Humpel, N., Owen, N., Iverson, D., Leslie, E., & Bauman, A. (2004). Perceived environment attributes, residential location, and walking for particular purposes. *American Journal of Preventive Medicine, 26*(2), 119-125.

Humpel, N., Owen, N., & Leslie, E. (2002). Environment factors associated with adults' participation in physical activity: A review. *American Journal of Preventive Medicine, 22*(3), 188-199.

Iwarsson, S. (2005). A long-term perspective on person-environment fit and ADL dependence among older Swedish adults. *Gerontologist, 45*(3), 327-336.

Jette, A. M., Davies, A. R., Cleary, P. D., Calkins, D. R. et al. (1986). The Functional Status Questionnaire: Reliability and validity when used in primary care. *Journal of General Internal Medicine, 1*(3), 143-149.

Kahana, E., Lovegreen, L., Kahana, B., & Kahana, M. (2003). Person, environment, and person-environment fit as influences on residential satisfaction of elders. *Environment and Behavior, 35*(3), 434-453.

Kaplan, S. (1995). The restorative benefits of nature: Toward an integrative framework. *Journal of Environmental Psychology, 15*(3), 169-182.

Keysor, J. J., & Jette, A. M. (2001). Have we oversold the benefit of late-life exercise? *Journal of Gerontology: Medical Sciences, 56A*(7), M412-M423.

Kuo, F. E., Sullivan, W. C., Coley, L., & Brunson, L. (1998). Fertile ground for community: Inner-city neighborhood common spaces. *American Journal of Community Psychology, 26*(6), 823-851.

Lawton, M. P. (1986). *Environment and aging* (2nd ed.). Albany, NY: Center for the Study of Aging.

Lawton, M. P. (1989). Environmental proactivity and affect in older people. In S. Spacapan & S.Oskamp (Eds.), *The social psychology of aging*. Newbury Park, CA: Sage.

Lawton, M. P., & Nahemow, L. (1973). Ecology and the aging process. In C. Eisdorfer & M. P. Lawton (Eds.), *The psychology of adult development and aging.* Washington DC: American Psychological Association.

Leyden, K. M. (2003). Social capital and the built environment: The importance of walkable neighborhoods. *American Journal of Public Health, 93*(9), 1546-1551.

Li, F., Fisher, K. J., Bauman, A., Ory, M. G., Chodzko-Zajiko, W., Harmer, P., Bosworth, M., & Cleveland, M. (2005). Neighborhood influences on physical activity in middle-aged and older adults: A multilevel perspective. *Journal of Aging and Physical Activity, 13*(1), 87-114.

Little, B. R. (1983). Personal projects: A rationale and method for investigation. *Environment and Behavior, 15*(3), 273-309.

Little, B. R. (2000). Persons, contexts, and personal projects: Assumptive themes of a methodological transactionalism. In S. Wapner, J. Demick, T. Yamamoto, & H. Minami (Eds.), *Theoretical perspectives in environment-behavior research: Underlying assumptions, research problems, and methodologies.* New York: Plenum.

Little, B. R., & Chambers, N. C. (2004). Personal project pursuit: On human doings and well-beings. In W. M. Cox & E. Klinger (Eds.), *Handbook of motivational counseling.* New York: John Wiley.

McAuley, E., Blissmer, B., Marquez, D. X., Jerome, G. J., Kramer, A. F., & Katula, J. (2000). Social relations, physical activity, and well-being in older adults. *Preventive Medicine, 31*(5), 608-617.

Metz, D. H. (2000). Mobility of older people and their quality of life. *Transport Policy, 7*(2), 149-152.

Mollenkopf, H., Marcellini, F., Ruoppila, I., Széman, Z., Tacken, M., & Wahl, H.-W. (2004). Social and behavioural science perspectives on out-of-home mobility in later life: Findings from the European project MOBILATE. *European Journal of Ageing, 1*(1), 45-53.

Morgan, K. (2003). Daytime activity and risk factor for late-life insomnia. *Journal of Sleep Research, 12*(3), 231-238.

Nezlek, J. B., Richardson, D. S., Green, L. R., & Schatten-Jones, E. C. (2002). Psychological well-being and day-to-day social interaction among older adults. *Personal Relationships, 9*(1), 57-71.

Omodei, M. M., & Wearing, A. (1990). Need satisfaction and involvement in personal projects: Toward an integrative model of subjective well-being. *Journal of Personality and Social Psychology, 59*(4), 762-769.

Pikora, T. J., Bull, F. C., Jamrozik, K., Knuiman, M. et al. (2002). Developing a reliable audit instrument to measure the physical environment for physical activity. *American Journal of Preventive Medicine, 23*(3), 187-194.

Saelens, B. E., Sallis, J. F., Black, J. B., & Chen, D. (2003). Neighborhood-based differences in physical activity: An environment scale evaluation. *American Journal of Public Health, 93*(9), 1552-1558.

Satariano, W. A., & McAuley, E. (2003). Promoting physical activity among older adults: From ecology to the individual. *American Journal of Preventive Medicine, 25*(3Sii), 184-192.

Schutzer, K. A., & Graves, B. S. (2004). Barriers and motivations to exercise in older adults. *Preventive Medicine, 39*(5), 1056-1061.

Shumway-Cook, A., Patla, A., Stewart, A., Ferrucci, L., Ciol, M. A., & Guralnik, J. M. (2003). Environmental components of mobility disability in community-living older persons. *Journal of the American Geriatrics Society, 51*(3), 393-398.

Silverstein, M., & Parker, M. G. (2002). Leisure activities and quality of life among the oldest old in Sweden. *Research on Aging, 24*(5), 528-547.

Singh, M. A. (2002). Exercise comes of age: Rationale and recommendations for a geriatric exercise prescription. *Journal of Gerontology: Medical Sciences, 57A*(5), M262-M282.

Skelton, D. A. (2001). Effects of physical activity on postural stability. *Age and Ageing, 30*(S4), 33-39.

Smith, J., & Baltes, P. B. (1997). Profiles of psychological functioning in the old and oldest old. *Psychology and Aging, 12*(3), 458-472.

Strawbridge, W. J., Deleger, S., Roberts, R. E., & Kaplan, G. A. (2002). Physical activity reduces the risk of subsequent depression for older adults. *American Journal of Epidemiology, 156*(4), 328-334.

Takano, T., Nakamura, K., & Watanabe, M. (2002). Urban residential environments and senior citizens' longevity in mega city areas: The importance of walkable green spaces. *Journal of Epidemiology and Community Health, 56*(12), 913-918.

Trost, S. G., Owen, N., Bauman, A. E., Sallis, J. F., & Brown, W. (2002). Correlates of adults' participation in physical activity: Review and update. *Medicine and Science in Sports and Exercise, 34*(12), 1996-2001.

Ulrich, R. S., Simons, R. F., Losito, B. D., Fiorito, E. et al. (1991). Stress recovery during exposure to natural and urban environments. *Journal of Environmental Psychology, 11*(3), 201-230.

Wahl, H. W., & Weisman, G. D. (2003). Environmental gerontology at the beginning of the new millennium: Reflections on its historical, empirical, and theoretical development. *Gerontologist, 43*(5), 616-627.

Wallenius, M. (1999). Personal projects in everyday places: Perceived supportiveness of the environment and psychological well-being. *Journal of Environmental Psychology, 19*(2), 131-143.

Wang, H.-X., Karp, A., Winblad, B., & Fratiglioni, L. (2002). Late-life engagement in social and leisure activities is associated with a decreased risk of dementia: A longitudinal study from the Kungsholmen Project. *American Journal of Epidemiology, 155*(12), 1081-1087.

Weuve, J., Kang, J. H., Manson, J. E., Breteler, M. M. B., Ware, J. H., & Grodstein, F. (2004). Physical activity, including walking, and cognitive function in older women. *Journal of the American Medical Association, 292*(12), 1454-1461.

World Health Organization (2003). *Health and development through physical activity and sport.* Geneva, Switzerland: World Health Organization.

The Effect of Viewing a Landscape on Physiological Health of Elderly Women

Joyce W. Tang
Robert D. Brown

SUMMARY. A quasi-experiment was undertaken to measure physiological characteristics of elderly women as they viewed different landscapes. Blood pressure and heart rate were monitored as elderly women living in a retirement centre viewed a natural landscape, a built landscape, and a control room with no view to the outside. Other characteristics of the individuals and the settings that have been shown to affect blood pressure and heart rate were controlled. The results indicated that, in all cases, viewing the natural landscape resulted in lower systolic and diastolic blood pressures and lower heart rates than those measured in the control room. Viewing the built landscape also had the general effect of lowering blood pressures and heart rates, but the effect was less consistent and the magnitude was smaller than that caused by the natural landscape. Lowering of blood pressure and heart rate have both been shown to be positively correlated with increased health and well-being, indicating the benefit of simply viewing a landscape. These results have

Joyce W. Tang, MLA, is in private practice in Calgary, Alberta.

Robert D. Brown, MLA, PhD, is Professor of Landscape Architecture, School of Environmental Design and Rural Development, University of Guelph, Guelph, Ontario, Canada N1G 2W1 (E-mail: rbrown@uoguelph.ca).

Address correspondence to: Robert D. Brown, School of Environmental Design and Rural Development, University of Guelph, Guelph, Ontario N1G 2W1, Canada (E-mail: rbrown@uoguelph.ca).

[Haworth co-indexing entry note]: "The Effect of Viewing a Landscape on Physiological Health of Elderly Women." Tang, Joyce W., and Robert D. Brown. Co-published simultaneously in *Journal of Housing for the Elderly* (The Haworth Press, Inc.) Vol. 19, No. 3/4, 2005, pp. 187-202; and: *The Role of the Outdoors in Residential Environments for Aging* (ed: Susan Rodiek, and Benyamin Schwarz) The Haworth Press, Inc., 2005, pp. 187-202. Single or multiple copies of this article are available for a fee from The Haworth Document Delivery Service [1-800-HAWORTH, 9:00 a.m. - 5:00 p.m. (EST). E-mail address: docdelivery@haworthpress.com].

Available online at http://www.haworthpress.com/web/JHE
© 2005 by The Haworth Press, Inc. All rights reserved.
doi:10.1300/J081v19n03_10

important implications for design of housing for the elderly. Even if individuals are unable or unwilling to go outside, they can still benefit from seeing out into a landscape. *[Article copies available for a fee from The Haworth Document Delivery Service: 1-800-HAWORTH. E-mail address: <docdelivery@haworthpress.com> Website: <http://www.HaworthPress.com> © 2005 by The Haworth Press, Inc. All rights reserved.]*

KEYWORDS. Landscape architecture, blood pressure, heart rate, therapeutic, attention restoration theory, human health

INTRODUCTION

There is a growing body of literature identifying the beneficial effects that natural landscapes can have on people. Studies have indicated that people are drawn to and prefer natural landscapes for viewing, driving, recreational activity, therapy, increasing productivity, promoting rehabilitation, and as a place to relax (Ulrich, 1981; Kaplan and Kaplan, 1989; Hartig et al., 1991; Cimprich, 1993; Pigram, 1993; Purcell et al., 1994; Cooper Marcus and Barnes, 1999; Ulrich, 1999).

The simple act of viewing a natural landscape seems to have a restorative effect on people. Post-operative patients have been found to recover more quickly, have fewer minor post-surgical complications, and require less potent narcotic drugs when in a room with a view of a wooded area compared to patients in a room with a view of a brick wall (Ulrich, 1984). During commuting to and from work subjects' blood pressures have been measured to be lower when they viewed natural landscapes than when they viewed built landscapes with concrete, buildings, and billboards (Parsons et al., 1998), and access to a view of the natural landscape was determined to reduce stress (Ulrich, 1999). However, support for facilities such as healing gardens in health care facilities has been minimal, and the problem has been identified as a lack of empirical data available for the decision makers and health care administrators (Hartig et al., 1991).

Theories and studies suggest there is a positive correlation between natural landscapes, and health and well-being. The *attention restoration theory* (James, 1892) suggests that exposure to natural landscapes provides people with a restorative environment that benefits health and well-being. Stemming from this theory are two streams of thought as to why landscapes provide restoration for people: *recovery of attention*

(Kaplan and Talbot, 1983); and *recovery from stress* (Ulrich, 1983; Kaplan, 1995).

In more recent years immerging research have started to explore the benefits of natural landscape in the form of outdoor gardens on the health outcomes of the elderly population as well as develop objective methods to generate more empirical data. These data suggest that the implied benefit of the outdoor garden or natural landscape on the elderly population is a predictable outcome. For example, one study found that stress levels were reduced when the elderly subjects performed the same activity in an outdoor garden versus that in an indoor environment (Rodiek, 2002). In this study, salivary coritsol was used as a measure of change in stress levels. A second study assessed blood pressure, heart rate, and "powers of concentration" of the elderly while being in an outdoor garden (Ottosson and Grahn, 2005). The results of this study found that the outdoor garden improved the level of concentration of the elderly subjects, thus suggesting that the outdoor garden could improve the performance of activities of daily living (Ottosson and Grahn, 2005).

In this paper we have further investigated the attention restoration theory by examining the effects of viewing natural and built landscapes on the physiological characteristics of blood pressure and heart rate, and on the psychological characteristic of mood. The effects were investigated through a quasi-experimental design utilizing a pretest-posttest comparison (Babbie, 1998). While many studies in the past have focused on populations of university students (Ulrich, 1999), with the advancing age of world populations we chose to study elderly subjects. The results of this study should be beneficial in the design of housing for the elderly.

METHOD

Instruments

Blood Pressure and Heart Rate

Measurements of heart rate and blood pressure are common in research and are simple, non-invasive, and do not disturb the subject. Blood pressure is typically recorded as a fraction, with the numerator being the systolic and the denominator the diastolic value. The systolic pressure occurs during ventrical contraction, and the diastolic pressure is the elastic recoil of the walls of the large arteries with forward propulsion of the blood during ventrical relaxation (Berne and Levy, 1993).

The Omron Wrist Blood Pressure Monitor with IntelliSense, Model HEM-609V was used throughout the experiment. It was calibrated against a research laboratory standard and was determined to be within 4 mm Hg for both systolic and diastolic pressures in all cases. Our study was not concerned with absolute values of blood pressure, but rather the relative changes due to experimental stimuli. To ensure that precise measurements were recorded we used the same instrument in every test, and took 5 measurements at each time. A research physiologist reviewed the data to identify any false readings. Remaining values were averaged to provide the reading.

The Omron is worn by wrapping it around the left wrist and using a velcro strip to attach it to itself and hold it in place. When the 'start' button is pressed, the instrument inflates applying pressure to the veins in the wrist, simultaneously measuring both heart rate and blood pressure in a single inflation. During each reading subjects are required to remain still, in an upright seated position with their left arm across their chest and their left hand on their right shoulder, and their left elbow supported by their right hand (see Figure 1).

FIGURE 1. Seating Position of Subject During the Collection of Blood Pressure Data

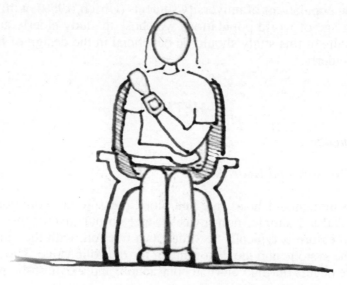

Mood Survey and Behavioural Information

Each subject's mood was measured through the use of the Profile of Moods Survey (POMS) short test that has been tested on many different groups varying in age, health, and educational background, and has been shown to be easily applied (McNair et al., 1992). The POMS has been applied in a wide variety of situations, thus providing a large amount of empirical support and a basis for comparison. Subjects were asked to complete the POMS short form during the pre-tests, and after each post-test.

A diary sheet was used during the pre-tests and post-tests to record subject behavioural information that research suggest are factors that affect blood pressure and heart rate, such as food or drink recently consumed, and any disruptions or interruptions that the subject perceived during the procedure.

PROCEDURE

All factors known to affect blood pressure were held constant, except *environment* which became the study's independent variable. A staff member of the retirement facility selected subjects with similar medical history, age, cultural group, socio-economic group, familiarity with the sites, travel distance, and care requirements (Dressler et al., 1986, Hartig et al., 1991; Lane et al., 1998; Jacob et al., 1989) thus increasing the validity and confidence of the results (Neuman; 1994). A total of 20 suitable subjects were selected. When these individuals were approached by the investigators and invited to participate, only 7 volunteered to do so. They were all female residents of a single retirement facility who were between the ages of 77 and 89 years old, with a mean age of 84, from a similar socio-economic setting, similar familiarity with the site, same travel distance (none), similar general health, and the same occupation (retired). All had completed high school, and two had attended a technical business college. All the subjects lived in Kingsway Retirement Centre in Toronto, were Caucasian, and had the same cognitive and mobility skills, and were from the same care group. No other information on any subject's history was known by the investigators.

Changes in air temperature as small as 5°C have been shown to elevate blood pressure levels (McLean et al., 1992) so the study was conducted indoors where the climate was carefully controlled. Seating position and walking distance have been shown to affect blood pressure

(Cottington et al., 1986; Hartig et al., 1991) so for each view the subject traveled the same distance and was seated in the same armchair and position. Ulrich et al. (1991) found that after viewing a natural setting, restoration measured by blood pressure was achieved in 4 to 7 minutes. Based on this result we decided to measure blood pressure and heart rate at both 5 minutes and 10 minutes after introducing each view.

All subjects were escorted and measured by the same investigator throughout the course of the study to reduce investigator variation. To minimize the influence of experimenter-subject contact the subject kept the wrist blood pressure monitor on the whole time. Throughout the procedure the subject was asked to not talk with anyone and to focus on the views.

During the entire procedure the investigator only spoke to a subject when it was necessary, by answering questions related to the study or giving instructions in short and simple sentences. Physical contact was minimal, and only occurred when the instrument was placed on each subject's arm, to start and stop the monitor during data collection, and removal of the instrument at the end of the experiment.

All readings for each subject were taken in the same seated position and on the same day in one continuous time frame, ensuring no day-to-day variation or other activities, such a eating or drinking, that might have affected readings. Factors such as mood and activity prior to the beginning of the test were not controlled, and were recorded in a diary.

All procedures were administered during daylight hours between 9:00 a.m. and 5:00 p.m. on April 5, 10, and 11, 2001. The subjects were asked ahead of time to wear loose-fitting, comfortable clothing on the day of the experiment. A female university student administered the experiment and took all the measurements. She treated every subject in the same pleasant manner.

At the beginning of each test, the program coordinator from the retirement facility approached the subject, introduced the investigator and the study, and assured the subject that the centre was in support of the project. This was intended to reduce any anxiety that the subject might have had about the study, allow them to become more familiar with the investigator, and to ask any questions they might have had.

The subject was asked to sit in an armchair in the control room, and while they relaxed the investigator provided them with a general description of the experimental procedure and familiarized them with the survey, the diary sheet, and the Omron instrument. Subjects were given no information about how to interpret any of the data, and were not told that the Omron measured their blood pressure and heart rate. After a two-minute rest period, pretest measurements were taken and recorded. This consisted of the investigator taking 5 consecutive blood pressure and heart rate measurements, and the subject completing the POMS test and diary sheet. When the pretest readings were complete, the experimenter escorted the subject to the appropriate site to view either the natural or the built landscape (randomly assigned).

After a short walk, the subject was seated in an armchair of the same style as that in the control room, and was asked to look out the window at the landscape. Blood pressure and heart rate measurements were taken after 5 minutes and after 10 minutes of viewing the landscape. The subject was then asked to again complete the POMS and diary sheet. Upon completion the subject was then escorted to the control room and the procedure was repeated identically for the other landscape view.

Study Area

The study was conducted at the Kingsway Retirement Residence for seniors in Toronto, Ontario. The pre-test measurements were taken in the control environment, a room with no outside views that was an equal distance from both the natural landscape stimulus and the built landscape stimulus (see Figure 2). This room was familiar to all the subjects, and was an area in which little or no activity was conducted. The furniture was not altered from its normal arrangement, except to add one comfortable armchair. All pretest measurements were taken in this location, with the subject facing a blank wall (see Figure 3).

The natural landscape view was provided from the activity room, which was a quiet, low-traffic area. The same type of armchair used in the control environment was placed in front of a large window looking directly onto the natural landscape. The natural landscape, for the purpose of this study, was defined as an environment that is predominantly vegetation, park-like in setting, and has minimal human induced change such as buildings and roadways (see Figure 4).

The built landscape was viewed from a large window in the parlour, a quiet, low-traffic area similar to the activity room. Each subject was again seated in a comfortable armchair in front of the window and asked to look directly out the window at a four-lane road. There was a narrow sidewalk in the foreground, which met directly with the road that was in the middle ground, with a plaza in the background (see Figure 5).

FIGURE 2. Layout of Centre Showing the Relative Distance Between Control Room and Stimulus Areas

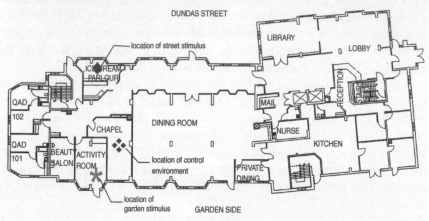

FIGURE 3. Control Room Setting

FIGURE 4. Stimulus A: Natural Landscape View–View of the Garden from Inside the Retirement Centre

RESULTS

On the day of the tests, one individual was found to be physically unable to wear the Omron instrument or to sit in the position necessary for taking measurements, and had to be removed from the study. Another individual was confused by the explanation of the procedure, and de-

FIGURE 5. Stimulus B: Built Landscape–Street View from Inside the Retirement Centre

cided to withdraw from the study. The participation rate was therefore 5 subjects, or 25% of the pool identified by the program coordinator.

In every case when a subject had viewed the *natural* landscape for 10 minutes their blood pressure and heart rate measurements were lower than in the pretest. The average reductions were 6.8 mmHg for systolic, 3.0 mmHg for diastolic, and 3.1 beats/minute for heart rate (see Table 1). The results for subjects viewing the *built* landscape for 10 minutes were less clear, with three subjects experiencing a reduction in blood pressure and heart rate, and two experiencing reductions in some measures but increases in others. However, the average effect of viewing the built landscape was a reduction of 5.2 mmHg for systolic, 1.6 mmHg for diastolic, and 1.3 beats/minute for heart rate. The differences between the effect of the two views indicated that participation in the study and the attention from the investigator was not the cause of changes in the blood pressure and heart rate of the subjects.

The results of statistical analysis found that the coefficient of variance values ranged from 11.21% to 0.7% with an average coefficient of variance value of 3.5%. These low values indicate a precise and reliable set of data and coincide with the calibration of the instrument to a hospital automatic blood pressure cuff. Therefore, the average scores that were used represent a good sample of blood pressure and heart rate readings for the entire subject population.

TABLE 1. The Stimulus Effect on the Average Change of the Entire Sample on Physical Health Measures

	Average Change of the Entire Sample			
	Time 1 Garden	Street	Time 2 Garden	Street
Systolic (mmHg)	**4.6**	2.3	**6.8**	5.2
Diastolic (mmHg)	**1**	0.7	**3**	1.6
Heart rate (pulse/minute)	**1.7**	0.3	**3.1**	1.3

The numbers indicate the average reduction in blood pressure levels and heart rate.

The sample size, although small, is not unusual in studies of elderly individuals. Takeuchi et al. (1991) studied the blood pressure of six elderly women, while Angeli et al. (1991) reported on the blood pressures of eight responders. Other studies have reported results of studies with sample sizes of less than ten individuals (e.g., Frontera et al., 2000; Morales et al., 1998).

The health effects of reducing blood pressure are well recognized, and quantification of these effects is being actively investigated. In a study of adults, Cook et al. (1995) found that a 2 mm Hg reduction in diastolic blood pressure (similar in magnitude to this study) would result in a 17% decrease in the prevalence of hypertension and a 15% reduction in risk of strokes. These values are only slightly less than those achieved through medical treatment. Similarly, a recent study (Adler et al., 2000) found that a 10 mm Hg decrease in systolic blood pressure (slightly larger than the values in this study) resulted in a 12% reduction in risk of complications related to diabetes.

Despite the small sample size, we investigated possible effects of several variables. The order in which the landscapes were viewed did not affect the results, with the natural landscape having a greater positive effect than the built landscape in every case. The results provided weak evidence that subjects who had higher blood pressure at the start of the study, had gardened in their past, and had lived in the centre a longer time, were more likely to experience an increasing effect over time, and were more likely to show a positive effect of viewing the built environment (see Table 2). Conversely, subjects who had lower blood pressure at the start of the study, had not gardened in their past, and had lived at the centre a shorter length of time, showed a greater positive effect of

TABLE 2. Those that Gardened in the Past–The Stimulus Effect on the Average Change of Physical Health Measures

	Gardeners-Average Change			
	Time 1 Garden	Street	Time 2 Garden	Street
Systolic (mmHg)	1.3	**6.2**	7.7	**9.1**
Diastolic (mmHg)	−1	**3.4**	**3.4**	2.5
Heart rate (pulse/minute)	**1.7**	−1.8	**3.7**	−0.4

the natural landscape, and almost no effect from viewing the built landscape.

The length of time that subjects viewed the landscape affected the results. Ulrich et al. (1991) found that viewing a natural landscape had a positive impact within the first four minutes. The results of our study supported this, with an indication of a positive effect within 5 minutes. However, by requiring subjects to view the landscapes for an additional 5 minutes we found an increase in the average stimulus effect of 49% for systolic pressure, 210% for diastolic pressure, and 81% for heart rate. This indicates that it would be worthwhile testing the effect over an even longer period of time to determine whether there would be an optimum amount of viewing time, after which the effect plateaus or drops off.

The POMS test was negatively received by the subjects. They found the form hard to read. Even when the form was read aloud to them, completing the survey four times during the study was found by participants to be tedious and redundant. Subjects commented "Oh, not this survey again." and "Is this the last time I have to fill this out?" The questions on the survey were also felt to be inappropriate. Subjects reported that the adjectives of mood in the POMS did not relate to them and their situation.

Based on these difficulties the results of the POMS short test were discarded. However, comments by the participants indicated that they did not consider themselves to be stressed. One subject said "I am not tense or anxious about anything. I am very relaxed." Another subject commented "I am always relaxed and calm."

Unsolicited comments also provided some insight into the effect on viewing landscapes on general mood. One participant said that after viewing the natural landscape she felt "very energetic and ready to be active," but after viewing the built landscape she said that she felt "a bit tired." Another started the study in an unpleasant and grumpy mood, but after viewing the natural landscape she became more pleasant and approachable.

CONCLUSIONS

This study found that a view to the outside, either to a natural or built landscape, lowered the blood pressure and heart rate of the elderly women in the study. The view of a natural landscape provided a much more consistent and more pronounced effect. The order in which the landscapes were viewed did not appear to affect the results.

Every participant in the study experienced a lowering of their blood pressure and heart rate when they viewed a natural landscape. The effect was evident after 5 minutes of viewing, and more pronounced after an additional 5 minutes. Viewing the built landscape resulted in a general lowering of both blood pressure and heart rate, but was less consistent across participants, and was lower in magnitude than the effect of the natural landscape. Two of the subjects experienced increases in some variables and decreases in others.

Preliminary analysis of the small sample size suggested that the effect was a function of the blood pressure level at the start of the study, whether or not the subject had gardened in their past, and how long they had lived at the retirement facility. The positive relationship between gardens and the health of the elderly provides some support to the restorative powers of the natural landscapes.

DISCUSSION

Further research similar to this on a larger population group, and different demographics and setting, could provide more insight as to the connection between nature and health. This information would encourage people to become more aware and supportive of creating the best environment to nurture physical and mental health.

During this study we encountered many difficulties and challenging situations. This might help to explain why so few studies of this nature have been done with elderly populations. We invited 15 retirement centres to be involved in the study but only 2 were willing to participate. Many explanations were offered as to why they didn't want to be involved, but most related to the perception that such as study would be quite 'disruptive' for a population on a daily routine.

We originally considered using elderly males in the study, but there was not a single individual in either of the two interested centres that wanted to be involved. Although there were considerably more women than men in the centres, we still experienced a very low participation rate.

Time and scheduling limitations required that the study be conducted in early April, a time when the natural landscape in Toronto did not yet have deciduous leaves or flowers. It would be valuable to test whether the effect would be different during summer or winter months.

Due to the small sample size, this study can only be seen as a pilot study that provided preliminary results. The findings, although strong and consistent, cannot be the basis for generalizations about the effects of viewing landscapes on the physiological health of elderly. They do provide considerable impetus for future studies to explore the relationship in more detail.

It is important that more studies be conducted to answer some central questions. Can a landscape be designed to have a maximum positive effect on health? Are there components of the population, such as people with health problems, who benefit the most from views of natural landscapes? Are there elements of the landscape that have the most positive effect? There are many questions that still need to be answered, but there is sufficient evidence for retirement home administrators to include a natural landscape as part of the environment of their facilities for enjoyment and promotion of good health.

The results of this study have important implications for design of housing for the elderly. Views of natural landscapes should be considered important for every room of a retirement facility. Even if individuals are unable or unwilling to go outside, they can still benefit from seeing out into a landscape.

REFERENCES

Angeli, P., M. Chiesa, L. Caregaro, C. Merkel, D. Sacerdoti, M. Rondana, and A. Gatta. (1991). Comparison of sublingual captopril and nifedipine in immediate treatment of hypertensive emergencies–a randomized, single-blind clinical trial. *Archives of Internal Medicine* 151(4), 678-682.

Babbie, E. (1998). *The Practice of Social Research, 8th Edition*. Toronto: Wadsworth Publishing Company.

Berne, R. M. and Levy, M. N. (eds.). (1993). *Physiology, 3rd Edition*. Toronto: Mosby Year Book.

Cimprich, B. (1993). Development of an intervention to restore attention in cancer patients. *Cancer Nursing* 16, 83-92.

Cook, N.R., J. Cohen, P.R. Hebert, J. O. Taylor, and C.H. Hennekens. (1995). Implications of small reductions in diastolic blood pressure for primary prevention. *Archives of Internal Medicine* 155(7), 701-709.

Cooper Marcus, C. and Barnes, M. (eds.). (1999). *Healing Gardens: Therapeutic Benefits and Design Recommendations*. New York: John Wiley & Sons.

Cottington, E. M., Matthews, K. A., Talbott, E., and Kuller, L. H. (1986). Occupational stress, suppressed anger, and hypertension. *Psychosomatic Medicine* 48, 249-260.

Dressler, W.W., Santos, J. E. D., and Viteri, F. E. (1986). Blood pressure, ethnicity, and psychosocial resources. *Psychosomatic Medicine* 48,509-519.

Frontera, W.R., V.A.Hughes, R.A. Fielding, M.A. Fiatarone, W.J. Evans, and R. Roubenoff. (2000). Aging of skeletal muscle: A 12-year longitudinal study. *Journal of Applied Physiology*. 88, 1321-1326.

Hartig, T., Mang, M., and Evans, G. W. (1991). Restorative effects of natural environment experiences. *Environment and Behavior.* 23, 3-26.

Jacob, R. G., Simons, A.D., Manuck, S. B., Rohay, J. M., Waldstein, S., and Gatsonis, C. (1989). The circular mood scale: A new technique of measuring ambulatory mood. *Journal of Psychopathology and Behavioral Assessment* 11, 153-173.

James, W. (1892). *Psychology: The Briefer Course*. New York, Holt.

Kaplan, R. and Kaplan, S. (1989). *The Experience of Nature*. New York: Cambridge, University Press.

Kaplan, S. (1995). The restorative benefits of nature: Toward an integrative framework. *Journal of Environmental Psychology* 15,169-182.

Kaplan, S. and J.F. Talbot. (1983). Psychological benefits of a wilderness experience. In I. Altman and J.F. Wohlwill (eds). *Behavior and the Natural Environment*. New York: McGraw-Hill.

Katzev, R. (1992). The impact of energy-efficient office lighting strategies on employee satisfaction and productivity. *Environment and Behavior* 24, 759-777.

Lane, J.D., Phillips-Bute, B. G., and Pieper, C. F. (1998). Caffeine raises blood pressure at work. *Psychosomatic Medicine* 60, 327-330.

McLean, J. K., Sathasivam, P., MacHaughton, K., and Graham, T. E., (1992). Cardiovascular and norepinephrine responses of men and women to two cold pressor test. *Canadian Journal of Physiological Pharmacology* 40, 36-42.

McNair, D. M., Lorr, M., and Doppleman, L. F. (1992). *Profile of Mood States, Revised 1992*. San Diego: Educational and Industrial Testing Service.

Morales, A.J., R.H. Haubrich, J.Y. Hwang, H. Asakura, and S.S.C. Yen. (1998). The effect of six months treatment with a 100mg daily dose of dehydroepiandrosterone (DHEA) on circulating sex steroids, body composition and muscle strength in age-advancing men and women. *Clinical Endrocrinology* 49(4), 421-432.

Neuman, W. L. (1994). *Social Research Methods: Qualitative and Quantitative Approaches, 2nd Edition*. Toronto: Allyn and Bacon.

Ottosson, J., and Grahn, P. (2005). A comparison of leisure time spent in a garden with leisure time spent indoors: On measures of restoration in residents in geriatric care. *Landscape Research* 30(1), 23-55.

Parsons, R. (1991). The potential influences of environmental perception on human health. *Journal of Environmental Psychology* 11,1-23.

Parsons, R., Tassinary, L., Ulrich, R., Hebl, M., and Grossman-Alexander, M. (1998). The view from the road: Implications for stress recovery and immunization. *Journal of Environmental Psychology* 18,113-140.

Pigram, J. J. (1993). Pages 400-426 *in* Garling, T. and Golledge, R. G. (eds.). *Behavior and Environment: Psychological and Geographical Approaches.* New York: Elsevier Science Publishers B. V.

Purcell, A. T., Lamb, R.J., Peron, E. M., and Falchero, S. (1994). Preference or preferences for landscape. *Journal of Environmental Psychology* 14,195-209.

Rodiek, S. D. (2002). Influence of an outdoor garden on mood and stress in older persons. *Journal of Therapeutic Horticulture* 13, 25-37.

Takeuchi, K., Abe, K., Yasujima, M., Sato, M., Tanno, M., Sato, K., and Yoshinaga, K. (1991). No adverse effect of non-steroidal anti-inflammatory drugs, sulindac and diclofenac sodium, on blood pressure control with a calcium antagonist, nifedipine, in elderly hypertensive patients. *Tohoku Journal of Experimental Medicine*, 165 (3), 201-208.

Ulrich, R. S. (1981). Natural versus urban scenes some psychophysiological effects. *Environment and Behavior* 13, 523-556.

Ulrich, R. S. (1983). Aesthetic and affective response to natural environment. Pages 85-120 *in* Altman, R. and Wohlwill, J. F. (eds.). *Behavior and the Natural Environment.* New York: Plenum Press.

Ulrich, R. S. (1984). View through a window may influence recovery from surgery. *Science* 224, 420-421.

Ulrich, R. S. (1986). Human responses to vegetation and landscapes. *Landscape and Urban Planning* 13, 29-44.

Ulrich, R. S. (1999). Effects of gardens on health outcomes: Theory and research. *in* Cooper Marcus, C. and Barnes, M. (eds.). *Healing Gardens: Therapeutic Benefits and Design Recommendations.* New York: Wiley and Sons, Inc.

Ulrich, R. S., Simons, R. F., Losito, B. D., Fiorito, E., Miles, M.A., and Zelson, M. (1991). Stress recovery during exposure to natural and urban environments. *Journal of Environmental Psychology* 11, 201-230.

Zajonc. R.B. (1980). Feeling and thinking: Preferences need no inferences. *American Psychologist*, 35, 349-365.

Restorative Environments in Later Life: An Approach to Well-Being from the Perspective of Environmental Psychology

Massimiliano Scopelliti
Maria Vittoria Giuliani

SUMMARY. This study proposes an analysis of restorative experiences of 192 elderly persons living in Rome. Perceived restorativeness of natural and built environments was examined, considering the influence of the social context of restoration and the activities performed in the environment. Restorativeness emerges as a consequence of complex person-environment transactions, in which place-specific processes occur. Moreover, a different influence of social interaction in natural and built environments was clearly identified; conversely, activities performed in the environment did not show a significant effect on perceived restorativeness. Theoretical and practical implications are discussed. *[Article copies available for a fee from The Haworth Document Delivery Service: 1-800-HAWORTH. E-mail address: <docdelivery@*

Massimiliano Scopelliti, PhD is affiliated with the Institute of Cognitive Sciences and Technologies, Italian National Research Council (ISTC-CNR). Maria Vittoria Giuliani, PhD, is Senior Researcher, Institute of Cognitive Sciences and Technologies, Italian National Research Council (ISTC-CNR) (E-mail: vittoria.giuliani@istc.cnr.it).

Address correspondence to: Massimiliano Scopelliti, Institute of Cognitive Sciences and Technologies, Italian National Research Council (ISTC-CNR), Via Nomentana 56, 00161 Rome, Italy (E-mail: massimiliano.scopelliti@istc.cnr.it).

[Haworth co-indexing entry note]: "Restorative Environments in Later Life: An Approach to Well-Being from the Perspective of Environmental Psychology." Scopelliti, Massimiliano, and Maria Vittoria Giuliani. Co-published simultaneously in *Journal of Housing for the Elderly* (The Haworth Press, Inc.) Vol. 19, No. 3/4, 2005, pp. 203-226; and: *The Role of the Outdoors in Residential Environments for Aging* (ed: Susan Rodiek, and Benyamin Schwarz) The Haworth Press, Inc., 2005, pp. 203-226. Single or multiple copies of this article are available for a fee from The Haworth Document Delivery Service [1-800-HAWORTH, 9:00 a.m. - 5:00 p.m. (EST). E-mail address: docdelivery@haworthpress.com].

Available online at http://www.haworthpress.com/web/JHE
© 2005 by The Haworth Press, Inc. All rights reserved.
doi:10.1300/J081v19n03_11

haworthpress.com> Website: <http://www.HaworthPress.com> © 2005 by The Haworth Press, Inc. All rights reserved.]

KEYWORDS. Residential environment, restorative experiences, natural environments, built environments, social interaction, activities

INTRODUCTION

Housing quality can be a critical factor in influencing health and psychological well-being of elderly people (Evans, Kantrowitz & Eshelman, 2002; Krause, 1996). Thomson, Petticrew, and Douglas (2003) clearly highlight a variety of healthy outcomes which are associated with housing improvements. Oswald and Wahl (2004) propose a scheme in which theoretical models addressing the relationships between housing and well-being of the elderly are integrated, and empirical evidence–though rather fragmented–supporting their analysis is provided. The authors emphasize the importance of integrating different levels of analysis of person-environment transactions, ranging from the micro-(house) to the macro-(neighbourhood and community) environment, and of assessing both objective and subjective aspects of housing. At the micro-level of analysis, Gitlin (2003) stresses the relevance of developing stronger theoretical models of elderly people-home relationships, mainly focusing on the role of private living arrangements in a successful adaptation to losses imposed by aging processes. At the macro-level, literature on residential satisfaction of elderly people has shown the multiplicity of environmental features which can enhance the assessment of everyday environment quality, ranging from aesthetics to facilities, perceived order and safety, lack of pollution and crowding (Perez, Fernandez, Rivera & Abuin, 2001; Christensen & Carp, 1987).

A central role in understanding the relationships between elderly people and their everyday environment is played by the construct of *person-environment fit* (Kahana, Lovegreen, Kahana & Kahana, 2003). This concept refers to the congruence between people's needs and preferences and environmental features, which may fit or not what they expect and/or desire; the perception of fit is supposed to promote well-being, while a lack of congruence may engender adverse consequences. With respect to the residential environment, a key concern is to identify what are the specific requirements which are susceptible to meet elderly people's needs and to enhance well-being.

Among others, the role of social interaction in fostering elderly people's health has been widely underlined in literature (Avlund, Lund, Holstein & Due, 2004; Homén & Furukawa, 2002; Unger, McAvay, Bruce, Berkman & Seeman, 1999; Allard, Allaire, Leclerc & Langlois, 1995). Hence, one aspect of person-environment fit, which is closely related to well-being for the elderly, has to do with the possibility for residential environments to promote social contact and support (Kweon, Sullivan and Wiley, 1998). Kahana et al.'s model (2003) identifies different domains of residential environments which can be a source of person-environment congruence/incongruence, referring to both social and environmental characteristics.

The relationships among perceived residential quality, satisfaction of people's needs and health are often emphasized in literature. Wallenius (1999) highlights the role of the perceived supportiveness of the environment in realizing personal projects, thus influencing psychological well-being. Kaplan and Kaplan (2003) provide a detailed analysis of the environmental features which can help constitute a "supportive environment." The authors stress the relevance of several positive requirements of such environments, which have to be easy to understand and able to hold up people in finding their way; have to encourage exploration, be attractive and support social interaction. Moreover, supportive environments have to provide a respite from attentional requests of everyday settings, which often lead to stress, irritability, distractibility and lack of efficacy in performing activities. Along the same line, Hartig, Johansson, and Kylin (2003) propose a social ecological model of the relationships between the residential environment and health. The authors analyze the residence in terms of stress, environmental demands and possibilities for restoration, also considering the influence of social and gender roles in shaping the processes connecting the residential experience to its outcomes. The relevance of focusing on social roles and users' specific needs when analyzing person-environment congruence in also highlighted by Scopelliti and Giuliani (2004).

Understanding which characteristics of supportive environments can promote beneficial effects is an essential requirement in order to produce helpful guidelines for design, improvement, and management of everyday residential environments. The theoretical framework proposed by the attention restoration theory (ART) offers an interesting perspective to address the issue of health and well-being for the elderly with reference to residential environments. ART (Kaplan & Kaplan, 1989; Kaplan, 1995) holds that four environmental properties account for the restorative potential of settings: being-away, extent, fascination, compatibility. *Being-away* implies a change of scenery and/or experience from routine, pressures and obligations of everyday life. *Extent* refers to the properties of connectedness and scope, in that all environmental elements are merged together in a harmonic whole (*coherence*) which promises to engage one's mind going further in the environment (*scope*). *Fascination* is related to the capability of environments to involuntarily catch one's attention, not demanding mental effort. A clear distinction is made between *soft* and *hard* fascination. Soft fascination implies a moderate intensity and is allowed by aesthetically pleasant stimuli which do not preclude the possibility for reflection. Hard fascination refers to a very intense involvement, leaving little room for thinking and thus supporting restoration to a lesser extent. Finally, *compatibility* refers to the perceived fit between the characteristics of the environment and the individual's purposes and inclinations; that is to say, "the setting must fit what one is trying to do and what one would like to do" (Kaplan, 1995, p. 173). Natural environments were generally found to have a higher restorative potential than built ones (Laumann, Gärling & Stormark, 2001; Purcell, Peron & Berto, 2001) and wilderness experience, in particular, emerged as a powerful source of restoration (Kaplan & Talbot, 1983; Hammitt & Madden, 1989). However, there is empirical evidence showing the possibility for restorative outcomes to occur also in built environments (Kaplan, Bardwell and Slakter, 1993).

ART assumes that the more an environment is compatible, fascinating, extensive, and allows being away to another world, the more it will have a restorative potential. It should be noted that these properties are not immanent in the environment, but are perceived by users; that is to say, environmental qualities are not a matter of fact, but a matter of people-environment transactions. So it is likely that different users, with different needs, may have a different perception of the same environment.

Studies on restorative environments have mainly focused on the measure of environmental properties (Hartig, Korpela, Evans & Gärling, 1997; Laumann et al., 2001). A tool to measure the restorative quality of environments (*Perceived Restorativeness Scale*-PRS), specifically focusing on the constructs of being-away, extent, fascination, and compatibility, was developed, and its psychometric properties were tested. Results substantially confirm the dimensional structure of the PRS, even though further examination is undoubtedly needed with reference to some subscales (Hartig et al., 1997; Purcell et al., 2001). Research was also addressed to the relationships between restorativeness and other constructs, such as preference (Purcell et al., 2001; Van den Berg,

Koole & van der Wulp, 2003), and the restorative experiences of people at specific stages (children, adolescents, university students) of the lifespan (Korpela & Hartig, 1996; Korpela, 2002; Korpela, Kytta & Hartig, 2002). What is still lacking is a lifelong approach to restorative experiences, in which attention is drawn on both choices for restoration by people at different stages of the life course—and, presumably, with different needs—and on the change of these choices through the lifespan. In this respect, an exception is provided by Scopelliti and Giuliani (2004). The authors propose an evaluation of environmental restorativeness which is based on the analysis of restorative experiences of people at different stages of the life course, also considering gender differences, with respect to relevance of the restorative components, affective reactions and social interaction.

In particular, the role of social aspects in restorative experiences has recently emerged as a research topic. Staats and Hartig (2004) underline the different influence of social interaction on perceived restorativeness in natural and urban environments. Social interaction was found to increase the restorative potential of an urban environment; conversely, in a natural environment perceived restorativeness is higher when people are alone, unless the environment is perceived as unsafe. In this situation, social interaction enhances the restorative qualities of nature through its influence on perceived safety. Also studies on wilderness show that the relationships with other people are not fundamental in natural environments for restorative outcomes to occur (Hammit, 1984; Hammitt & Madden, 1989).

These findings raise the question of the role of social interaction in restorative experiences of elderly people. On the one hand, the influence of social interaction on elderly people's health has been repeatedly found in literature; on the other, there is empirical evidence showing the positive effect of solitude in highly restorative environments. Hence, understanding whether or not social interaction is an important factor in influencing the restorative value of environmental experiences for elderly people still remains an open issue. In particular, an issue deserving attention is whether the effect of social interaction is independent of the typology of setting (e.g., natural vs. built) or it can be considered as place-specific to some extent. Lastly, a main concern is on other "contextual variables" (e.g., specific use of environments) which can contribute to enhance the positive outcomes of elderly persons' transactions with the environment.

THE EMPIRICAL RESEARCH

General Aim

The aim of this study was to identify some environmental conditions which can enhance the potential benefits of place experiences for elderly persons living in an urban context. In this respect, we addressed the evaluation of the restorative potential of environments by adopting a place-specific perspective (Russell & Ward, 1982; Canter, 1986), in order to develop helpful guidelines to implement more healthy environments, in which the elderly can satisfy their needs.

A two-phase approach was developed. In the first qualitative phase, data were collected through in-depth interviews in order to identify the types of environment that elderly people consider as being restorative, both in natural and in built context. Such a procedure emphasizes the focus of this study, which is on personal experiences. Scott and Canter (1997) provide empirical evidence that people conceptualize places in a different way when they are asked to focus on environmental features vs. personal experience in an evaluation task through a slide presentation. In addition, we were interested in identifying some contextual variables which can contribute to increase the restorative potential of the experiences. In the second phase, a quantitative approach to the examination of the restorative potential of the environments which were most frequently mentioned in the interviewees' descriptions was applied. A preliminary aim of this phase was to verify the dimensional structure of the PRS. Then, the influence of the contextual variables which emerged in the first phase on the restorative value of the experiences was explored.

Qualitative Study

Method

We contacted 48 elderly people of both genders (aged from 60 to 85; $M = 70.54$; S.D. $= 6.78$) living in an urban setting, and asked them to list as many environments as possible in which they usually had restorative experiences. Respondents were not asked to focus merely on residential settings, but to think about restorative environments in general, in order to get broad-spectrum insights on the characteristics of recovering settings. We assumed that also outcomes from non-residential environments can provide useful information to enhance the restorative qualities of residential settings. Participants were also asked to explain the reasons why the experience in those environments was able to promote a general condition of well-being.

Results

Different environments were recognized as restorative. Among natural ones, mountains, urban parks, lakes, seaside and countryside; among built ones, museums, historical towns, pedestrian-only plazas in the city center, cinemas, restaurants, town squares during a feast. At least two contextual variables influencing the restorative potential of the experiences emerged, namely social interaction and activities performed in the environment. A clear distinction between a preference for social interaction with "relevant others" (partner, friend, relatives) or, alternatively, solitude was outlined for different environments; at the same time, passive/relaxing and active/dynamic behaviors were associated with different restorative experiences.

Quantitative Study

On the basis of the results of the first study, the four natural and the four built environments with the highest frequencies of indication were selected. The settings were a mountain, an urban park, a seaside, a countryside, a pedestrian-only plaza in the city center, a museum, an historical town, a town square during a feast. With reference to these environments, we explored the role of both social interaction and activities performed on the characterization of restorative experiences of elderly people. Gender differences were also taken into consideration.

Hypotheses

According to the literature on wilderness (Hammitt, 1984; Hammitt & Madden, 1989), it was hypothesized that solitude increases the perceived restorativeness in natural environments; conversely, social interaction was hypothesized to enhance the restorative potential of built settings.

No hypotheses were advanced with reference to the influence of the kind of activity performed on the restorative value of natural and built environments. In this respect, a highly place-specific pattern of experience was expected.

Method

Design

A 2 (gender) × 2 (social interaction) × 2 (typology of activity) design was used. We manipulated the variables "social interaction" (solitude vs. presence of social interaction) and "typology of activity" (passive vs. active behaviors).

The dependent variables we measured were the restorative properties of environments, namely, being-away, extent, fascination, and compatibility.

Stimuli

For each selected environment a brief narrative was created. The descriptions included both static (fixed features) and dynamic (what was going on) environmental characteristics, which were derived from the respondents' reports. Our interest in real experiences and in an experimental manipulation of some variables led us to adopt the scenario methodology instead of a static slide presentation. In addition, the effectiveness of a scenario manipulation in the study of restorative experiences was shown by Herzog, Chen and Primeau (2002).

The environments were always described in the daylight and in good weather (except for the museum, which is an indoor environment). For each scenario, after a common description of the physical environment, the social and behavioral aspects of the situation were described, according to the experimental condition; namely, if the respondent was alone or with relevant others and if he/she was performing a dynamic vs. passive activity.

The "city center" scenario is presented as an example:

> You are in the pedestrian island of a urban center. It is a sunny day. You can see a lot of shops with beautiful windows on one side and smart cafes with elegant dining tables in the open on the other side. In the middle of the square there is an artistic fountain, and all around you can notice well-kept buildings in the same style. There are some people in the square, walking slowly, watching at the windows and going shopping. Other people are sitting in the open. The environment is lively. You can hear people talking softly and the sound of water from the fountain. (Common description)

You are enjoying your time with a person you love. You spend some time walking all around, watching the windows and all the beautiful buildings surrounding you. (Condition 1: social interaction and dynamic activities)

You are enjoying your time by yourself. You spend some time walking all around, watching the windows and all the beautiful buildings surrounding you. (Condition 2: solitude and dynamic activities)

You are enjoying your time with a person you love. You spend some time sitting in the open, having a drink, and chatting. (Condition 3: social interaction and passive activities)

You are enjoying your time by yourself. You spend some time sitting in the open, having a drink, and watching the beautiful buildings surrounding you. (Condition 4: solitude and passive activities)

Sample and Procedure

We contacted 192 elderly people living in Rome, stratified with respect to gender (96 males and 96 females). We opted to select mainly young elderly respondents (aged from 63 to 78; M = 68.23; S.D. = 4.38) in good health, in order to avoid that physical or cognitive impairments would affect a first-hand experience (accessibility and practicability of some settings) and a reliable evaluation of the environments. The majority of respondents (92%) were married. Participants were administered a questionnaire in which they were asked to evaluate the scenarios of the 8 restorative environments selected from the first study, according to the different experimental conditions. Subjects were asked to imagine themselves in each environment and to assess the quality of the environment by referring to their personal experience. Because we were interested in direct experience in the environments, respondents were asked to consider whether they were familiar with the proposed situation or not. Subjects considering even a single environment as unfamiliar were meant to be excluded. However, no exclusion was necessary, because all respondents judged the situations described as highly familiar.

In order to avoid cognitive fatigue, each questionnaire included only four environments (always 2 natural and 2 built environments), in a random order. The randomization allowed each environment to be evaluated by 96 respondents, well balanced as to gender and experimental conditions.

Tools

We employed an Italian version of the PRS (Hartig et al., 1997), consisting of 26 7-step (from 0 to 6) agree/disagree Likert-type items. The scale measures the constructs of being-away (6 items), extent (both coherence–4 items and scope–4 items), fascination (5 items), compatibility (7 items). The PRS was presented after the description of each scenario. At the end of the PRS, a single agree/disagree Likert-type item focusing on positive outcomes of the experience ("I really feel restored in this environment") was presented. A final section to collect socio-demographics was included.

Analyses

A factor analysis was performed for each environment in order to confirm the dimensional structure of the data and to identify the main components on which the perceived restorativeness of environments is based. Mean scores on the factors we extrapolated were computed for each environment; an overall score of restorativeness (the mean of the scores on extracted factors), was also calculated for each environment, in order to evaluate the global level of perceived restorative potential.

To compare the level of perceived restorativeness in the different environments a series of repeated measure analyses of variance (ANOVAs) was performed, in which the global scores of restorativeness were used.

A mixed design ANOVA was performed in order to explore the relationships between gender, social interaction, performed activities and the perception of the restorative potential of each of the eight environments considered in this study. Gender, social interaction and performed activities were between-groups factors; restorative components were within-groups factors. Age was included as a control variable in the analyses, because of the quite large range of this variable in our sample. The dependent variables used for the analyses were the mean scores of the components extrapolated in the factor analyses.

Post hoc comparisons were performed by using the Bonferroni correction, with p-level $< .05$. Finally, in order to evaluate the relationships between perceived restorativeness and psychological outcomes, a correlation analysis between the two variables was carried out for each environment.

Results

The factor analyses performed for the 8 environments showed a somewhat place-specific dimensional structure of the data (Table 1). Even though the structural properties of the PRS are generally confirmed, the four restorative components proved to have a different consistency across environments. The *seaside* emerged as a very peculiar environment, in that a combination of being-away and compatibility in the *Being-Away/No obstacles* factor and compatibility and fascination in the *Compatibility/Interest* factor were outlined. Nonetheless, the internal consistency of the factors extrapolated in each environment was always very high, Cronbach's α ranging from .79 to .94.

TABLE 1. Factor Analyses.

ENVIRONMENT		COMPONENT			
Urban Park	*Factors and Alpha*	Being-Away α = .91	Extent α = .93	Fascination α = .90	Compatibility α = .94
	N. of items	Ba: 6	Coh: 4 - Sc: 2	Fa: 5 - Sc 2	Com: 7
Mountain	*Factors and Alpha*	Being-Away α = .88	Extent α = .86	Fascination α = .82	Compatibility α = .87
	N. of items	Ba: 6	Coh: 4 - Sc: 4	Fa: 5	Com: 7
Seaside	*Factors and Alpha*	B-A/No obstacles α = .88	Coherence α = .88	Scope α = .79	Compat./Interest α = .93
	N. of items	Ba: 5 - Comp: 2	Coh: 4	Sc: 4 - Fasc: 1	Com: 5 - Fas: 4
Countryside	*Factors and Alpha*	Being-Away α = .90	Extent α = .91	Fascination α = .92	Compatibility α = .93
	N. of items	Ba: 5	Coh: 4 - Sc: 4	Fasc: 5 - Ba: 1	Com: 7
Urban plaza	*Factors and Alpha*	Being-Away α = .89	Extent α = .91	Fascinatio/ α = .94	Compatibility α = .90
	N. of items	Ba: 6	Coh: 4 - Sc: 2	Fa: 5 - Sc: 2	Com: 7
Museum	*Factors and Alpha*	Being-Away α = .83	Extent α = .86	Fascination α = .94	Compatibility α = .88
	N. of items	Ba: 6	Coh: 4 - Sc: 2	Fa: 5 - Sc: 2	Com: 7
Historical town	*Factors and Alpha*	Being-Away α = .91	Extent α = .07	Fascination α = .00	Compatibility α=.93
	N. of items	Ba: 6	Coh: 4 - Sc: 2	Fa: 5 - Sc: 2	Com: 7
Feast in a town square	*Factors and Alpha*	Being-Away α = .92	Extent α = .90	Fascination α = .92	Compatibility α = .91
	N. of items	Ba: 6	Coh: 4 - Sc: 2	Fa: 5 - Sc: 2	Com: 7

Note: Ba: Being-Away; Coh: Coherence; Sc: Scope; Fasc: Fascination; Com: Compatibility

Mean scores on the dimensions we extrapolated in the factor analyses were then calculated for each environment, and used in the subsequent analyses.

The repeated measure ANOVAs performed highlighted an interesting level of perceived restorativeness in the eight environments (Table 2). As a consequence of the procedure we adopted (each respondent randomly evaluating 4 of 8 environments), a 8-level repeated measures analysis was not allowed. However, result emerged as a consequence of a series of 2-level repeated measures analyses. Elderly people perceive the *countryside* and the *mountain* as the most restorative environments; the *urban plaza* and the *feast in a town square* turned out to be the least. Table 2 shows in detail the significant differences in the perceived restorativeness between environments.

The mixed design ANOVAs showed a place-specific influence of the design variables (gender, social interaction, typology of activity) on the restorative potential of the eight environments.

Among natural environments, in the *urban park* scenario a significant difference in the perception of the four restorative components emerged ($F_{(3, 264)} = 5.16$, p < .01); compatibility and being-away were found to score higher than fascination; coherence did not show a significant difference when compared to the other components (Table 3). No significant main effect of gender ($F_{(1, 88)} = 0.16$, n.s.), social interaction ($F_{(1, 88)} = 0.16$, n.s.), performed activity ($F_{(1, 88)} = 0.30$, n.s.), or interaction between any of these factors was found.

As regards the *mountain* scenario, a significant difference in the perception of the four restorative components was found ($F_{(3, 264)} =$

TABLE 2. Perceived Restorativeness of the 8 Environments

ENVIRONMENT	MEAN*	S. DEV.
Countryside	4.36[a]	.84
Mountain	4.32[a]	.74
Historical town	4.05[b]	.91
Urban park	3.59[c]	1.07
Museum	3.24[c,d]	.81
Seaside	3.24[c,d]	.93
Urban plaza	3.08[d,e]	1.12
Feast in a town square	2.87[e]	1.07

* the letters [a,b,c,d,e] refer to significant differences between environments

14.09, p < .001); extent and fascination were found to score higher than compatibility and being-away (Table 3).

A significant interaction effect between social interaction and component was also found ($F_{(3, 264)}$ = 13.45, p < .001); solitude significantly increases the perception of being-away and fascination (Figure 1). No significant main effect of gender ($F_{(1, 88)}$ = 0.04, n.s.), social interaction ($F_{(1, 88)}$ = 0.59, n.s.), performed activity ($F_{(1, 88)}$ = 0.01, n.s.), or interaction between any of these factors was found.

With respect to the *seaside* scenario, a significant difference in the perception of the four restorative components emerged ($F_{(3, 264)}$ = 4.64, p < .01); compatibility/interest was found to score higher than the other three components (Table 3). The interaction effect between social interaction and component showed a tendency to significance ($F_{(3, 264)}$ = 2.45, p <

TABLE 3. Natural Environments: Perception of Restorative Components

URBAN PARK			MOUNTAIN			SEASIDE			COUNTRYSIDE		
Factor	Mean*	S. dev.	Factor	Mean*	S. dev.	Factor	Mean*	S. dev.	Factor	Mean*	S. dev.
B-A	3.66a	1.24	B-A	4.10b	.98	BA/No	3.18b	1.05	B-A	4.22b	.99
COH	3.54a,b	1.17	EXT	4.57a	.87	COH	3.12b	1.27	EXT	4.31a,b	1.06
COM	3.78a	1.28	COM	4.12b	1.00	CO/IN	3.46a	1.13	COM	4.44a,b	1.13
FAS	3.36b	1.34	FAS	4.49a	.91	SCO	3.20b	.90	FAS	4.46a	.83

* the letters a,b refer to significant differences between environments
Note: B-A: being-away; BA/No: being-away/no obstacles; COH: coherence; COM: compatibility; CO/IN: compatibility/interest; FAS: fascination; EXT: extent; SCO: scope

FIGURE 1. Mountain: Social Interaction by Component

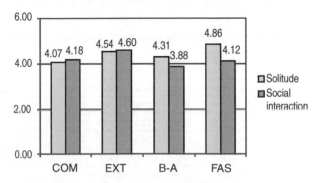

Note: COM: compatibility; EXT: extent; B-A: being-away; FAS: fascination

.08), people with relevant others scoring higher on compatibility than people alone. No significant main effect of gender ($F_{(1, 88)} = 0.02$, n.s.), social interaction ($F_{(1, 88)} = 0.24$, n.s.), performed activity ($F_{(1, 88)} = 0.01$, n.s.), or interaction between any of these factors was found.

As to the *countryside* scenario, a significant difference in the perception of the four restorative components emerged ($F_{(3, 264)} = 2.94$, p < .05); fascination was found to score higher than being-away (Table 3). The interaction effect between social interaction and component showed a tendency to significance ($F_{(3, 264)} = 2.51$, p < .08), people alone scoring higher on fascination and being-away than people with relevant others. No significant main effect of gender ($F_{(1, 88)} = 0.01$, n.s.), social interaction ($F_{(1, 88)} = 1.83$, n.s.), performed activity ($F_{(1, 88)} = 0.37$, n.s.), or interaction between any of these factors was found.

Among built environments, a significant difference in the perception of the four restorative components emerged in the *urban plaza* scenario ($F_{(3, 264)} = 5.25$, p < .01); fascination was found to score higher than both coherence and compatibility (Table 4).

A significant interaction effect between social interaction and component was also found ($F_{(3, 264)} = 3.49$, p < .05); social interaction significantly increases the perception of compatibility (Figure 2). No significant main effect of gender ($F_{(1, 88)} = 0.01$, n.s.), social interaction ($F_{(1, 88)} = 0.21$, n.s.), performed activity ($F_{(1, 88)} = 0.01$, n.s.), or interaction between any of these variables was found.

In the *museum* scenario, a significant difference in the perception of the four restorative components emerged ($F_{(3, 264)} = 57.84$, p < .001); coherence was found to score higher than the other three components, and the level of fascination showed to be higher than the level of both being-away and compatibility (Table 4). The effect of social interaction showed a tendency to significance ($F_{(1, 88)} = 3.71$, p < .08), people in the

TABLE 4. Built Environments: Perception of Restorative Components

URBAN PLAZA			MUSEUM			HISTORICAL TOWN			F. IN A TOWN SQUARE		
Factor	Mean*	S. dev.	Factor	Mean*	S. dev.	Factor	Mean*	S. dev.	Factor	Mean*	S. dev.
B-A	3.02[a,b]	1.48	B-A	2.80[c]	.99	B-A	4.08[a]	1.23	B-A	3.31[a]	1.37
COH	2.93[b]	1.13	COH	3.83[a]	.99	COH	4.15[a]	.97	COH	2.39[c]	1.37
COM	2.99[b]	1.37	COM	2.78[c]	.96	COM	3.78[b]	1.11	COM	2.94[b]	1.34
FAS	3.37[a]	1.34	FAS	3.55[b]	1.09	FAS	4.21[a]	.86	FAS	2.83[b]	1.49

* the letters [a,b,c] refer to significant differences between environments
Note: B-A: being-away; COH: coherence; COM: compatibility; FAS: fascination

condition of social interaction perceiving a higher level of restorativeness. A significant interaction effect between social interaction and component was also found ($F_{(3, 264)}$ = 5.80, p < .01); social interaction significantly increases the perception of compatibility and fascination (Figure 3). No significant main effect of gender ($F_{(1, 88)}$ = 0.11, n.s.) and performed activity ($F_{(1, 88)}$ = 0.01, n.s.), or interaction between these two variables was found.

In the *historical town* scenario, a significant difference in the perception of the four restorative components emerged ($F_{(3, 264)}$ = 9.53, p < .001); compatibility was found to score lower than the other three components, with no significant difference between being-away, coherence and fascination (Table 4). A significant interaction effect between so-

FIGURE 2. Urban Plaza: Social Interaction by Component

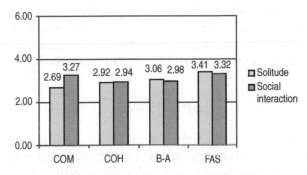

Note: COM: compatibility; COH: coherence; B-A: being-away; FAS: fascination

FIGURE 3. Museum: Social Interaction by Component

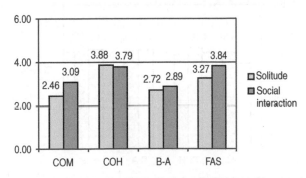

Note: COM: compatibility; COH: coherence; B-A: being-away; FAS: fascination

cial interaction and component was also found ($F_{(3, 264)} = 3.62$, $p < .05$); social interaction significantly increases the perception of compatibility (Figure 4). No significant main effect of gender ($F_{(1, 88)} = 0.05$, n.s.), social interaction ($F_{(1, 88)} = 0.53$, n.s.), performed activity ($F_{(1, 88)} = 0.70$, n.s.), or interaction between any of these variables was found.

With respect to the *feast in the town square* scenario, a significant difference in the perception of the four restorative components emerged ($F_{(3, 264)} = 13.19$, $p < .001$); being-away was found to score higher than the other three components, and the level of both compatibility and fascination showed to be higher than the level of coherence (Table 4). A significant interaction effect between social interaction and component was also found ($F_{(3, 264)} = 6.99$, $p < .05$); social interaction significantly increases the perception of being-away and compatibility (Figure 5). No significant main effect of gender ($F_{(1, 88)} = 0.16$, n.s.), social interaction ($F_{(1, 88)} = 2.26$, n.s.), performed activity ($F_{(1, 88)} = 0.03$, n.s.), or interaction between any of these variables was found.

The correlation analysis between perceived restorativeness and psychological outcomes was significant, showing a positive relationship between the two variables in all the 8 environments (Table 5).

DISCUSSION

A variety of results came out from this study. With respect to the perception of the restorative components by elderly people, an interesting pattern was outlined. Being-away and compatibility emerged as very

FIGURE 4. Historical Town: Social Interaction by Component

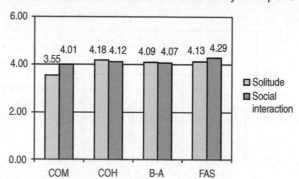

Note: COM: compatibility; COH: coherence; B-A: being-away; FAS: fascination

FIGURE 5. Feast in a Town Square: Social Interaction by Component

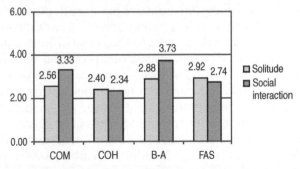

Note: COM: compatibility; COH: coherence; B-A: being-away; FAS: fascination

TABLE 5. Perceived Restorativeness and Psychological Outcomes: Correlation Analyses

Psychological outcome		Perceived restorativeness (overall score)							
		Urban park	Mountain	Seaside	Countryside	Urban plaza	Museum	Historical town	Feast in a town square
Restoration	r	.862	.853	.877	.893	.914	.881	.915	.834
	p	< .001	< .001	< .001	< .001	< .001	< .001	< .001	< .001

Note : N = 96

consistent dimensions across different environments; also fascination proved to be a consistent dimension, even though a couple of items from different restorative components (usually scope) often showed a high factor loading on this factor. It is worth nothing that scope items with a high factor loading on fascination refer to exploration, which is undoubtedly an activity with potential fascinating outcomes, suggesting that the theoretical distinction between fascination and scope is probably not so clear. Scope may represent an additional source of positive stimulation, going beyond a passive perception of pleasant/aesthetic elements, and requiring an active people-environment transaction. It would be interesting to further explore whether this hypothesis applies only to elderly people, or can it be consistent across different age groups. The component of extent was very often extrapolated, with the property of coherence being the most important dimension within the factor. In fact, beyond the four coherence items, also the two scope items with a high factor loading on this component imply the idea of harmony and relatedness in the environment. Scope emerged as the most problematic dimension, questioning the effective relevance of this component in elderly people's perception of environments. An interesting finding came out with respect to the seaside. Restorative properties emerged as a mixture of the dimensions proposed by ART. Taken together, these results call for a deep theoretical reflection on restorative components, in order to clearly define the distinction between some of them. In addition, some improvements of the tools which have been developed to measure restorativeness are probably needed.

The restorative potential of natural environments dramatically emerged, confirming previous findings (Kaplan & Talbot, 1983). An untouched nature (countryside and mountains) seems to be more easily associated with the idea of restoration, thus suggesting practical guidelines to improve the residential environment of elderly people. What can be important to them is not simply to have a natural spot in their residential environment, but probably the possibility to spend their time in touch with a nature with few man-made elements. In fact, the urban park emerged as a restorative environment, but not so restorative as the countryside and the mountain. In this respect, Herzog (1985) underlines the role of water in influencing preference; Horne, Boxall, and Adamowicz (2005) show that a variety of different plants, and a higher concern for scenic beauty can enhance preference for forest sites. Given the relationships between preference and restorativeness (Purcell et al., 2001; Van den Berg et al., 2003), the presence of water and a care for biodiversity could be easily taken into account in planning and improvement of residential parks, in order to promote more healthy environmental conditions.

Interestingly, the restorative potential of some built environments were also found. It strongly emerged with reference to the historical town and, to a lesser extent, to the museum (Kaplan et al., 1993). The urban plaza and the town square during a feast did not emerge as highly restorative environments, in contrast to the findings of the pilot study, in which many respondents included these environments among their indications. This discrepancy can be due to some shortcomings in the scenarios, which presumably were not able to focus on those environmental features which were considered in the first phase by respondents, when listing restorative environments. As to the historical town, this result supports the role of aesthetic/artistic value of places in promoting restoration for elderly persons, and has practical implications also for residential environments. Elderly people can benefit from the maintenance of order and good environmental conditions in the residential setting (Christensen & Carp, 1987); moreover, their well-being could be enhanced by improving the aesthetic value in the outdoors. An even more interesting proposal would be to merge together the beneficial influence of art and nature, for example, by coherently adding artistic elements in an urban park.

However, it is important to underline the positive association between the perception of restorative qualities and healthy psychological outcomes in both natural and built environments.

Another issue is related to the perception of restorative components in different settings. In natural environments (countryside and mountain) fascination and extent generally score higher than both compatibility and being-away, thus suggesting a clear direction through which restoration is achieved. The higher relevance of a mixture of compatibility and fascination at the seaside and both compatibility and being-away in the urban park shows that restoration in less pristine natural settings is much more dependent on a matter of person-environment fit (Kahana et al., 2003). In completely man-made environments (urban plaza, museum, historical town and feast in a town square) it is probably a lack of compatibility the main factor producing a decrease in perceived restorativeness. Elderly people seem to undergo a general difficulty with some settings in the residential environment, which might be the consequence of a lack of accessibility or a shortage of specific facilities capable to meet their needs. Taken together, these results suggest that restoration may occur in specific settings through different processes, in which different restorative components play a key role. In general, even though ART holds that all restorative properties are important for a deep restoration, our data show that it is possible to hypothesize that they are at least unequally relevant in different environments. This result confirms previous research (Herzog, Maguire & Nebel, 2003; Scopelliti & Giuliani, 2004).

Furthermore, it is important to analyze to what extent the specific person-environment transactions can influence the perceived restorativeness of experiences. The role of the two contextual variables we investigated (social interaction and performed activities) turned out to be very different.

As regards social interaction, it emerged to play an important and place-specific role in the perception of the restorative value of experiences. Among natural environments, solitude was found to increase the perceived restorativeness of both the mountain and the countryside, which undoubtedly show the lowest level of man-made elements, thus confirming previous research on wilderness (Hammitt, 1984; Hammitt & Madden, 1989). More specifically, being alone allows a better perception of the fascinating components of the environment, and a higher level of separation from everyday routine and obligations. In other words, other people seem to be a source of distraction from the beauty of the environment, and a reminder of usual practices. Green areas were found to promote social interaction in the residential environment (Kweon, Sullivan & Wiley, 1998), but our data suggest that in urban parks the presence of other people might have no effect on perceived restorativeness.

The influence of social interaction in built environments is in the opposite direction (Staats & Hartig, 2004). The presence of relevant others particularly increases one dimension of restorativeness, namely, compatibility, in all four environments; the influence on other components (fascination at the museum, being-away in the town square) is highly place-specific. An aspect to be considered is that built environments tend to "change" more rapidly than natural ones, so calling for a continuous adaptation which can be a difficult goal to achieve by elderly people alone. Social rules and environmental demands may be much more oppressive, and other people can be a source of comfort, even in familiar settings, against potential dangers (e.g., thieves, traffic, etc.). Conversely, natural settings are probably less susceptible to change and less demanding in terms of social rules; hence, the elderly can manage the environment simply relying on their acquired competence, and social interaction shows no influence on perceived compatibility. Staats and Hartig (2004) found that solitude increases the restorative potential of a natural environment when there is no perception of danger. It is possible to speculate that thinking about familiar natural experiences implicitly decrease the perception of potential risks in the environment, and perceived control is higher. An interesting result emerged from the seaside scenario, in that social interaction increases the perception of compatibility, similarly to the built environments. A possible explanation is that such an experience for elderly people in Italy has much more to do with a man-made environment (bathing establishments, crowding, etc.) than with a natural setting. On the whole, these findings substantially confirm our hypotheses on the different role of social interaction in natural and built environments.

With reference to activities performed in the environment, we did not find any significant effect on the restorative potential of experiences. Again, the difference between the importance of behaviors, which emerged in the first phase, and the results of the experimental manipulation is noticeable. A possible explanation can be found in the methodology we adopted. The scenario manipulation is presumably much more effective in displaying the social context of restoration (being alone or with other people) than the occurrence of preferred activities. In other words, the distinction between passive/relaxing and active/dynamic behaviors do not seem to be so important as the specific activity performed in the environment, with reference to which preferences are likely to be highly idiosyncratic.

Finally, no gender differences were found in the perception of the restorative potential of environments, confirming previous research (Scopelliti & Giuliani, 2004). However, gender differences might be outlined for people at earlier stages of the lifespan and, with respect to elderly people, the scenarios may be ineffective to focus on specific environmental features which are likely to be perceived in a different way by men and women. Further research on this topic is probably needed.

CONCLUSION

This article provides an investigation on restorative experiences of elderly people living in an urban context. Restorativeness emerged as a consequence of complex place experiences, in which environmental, social and behavioral components interact within specific settings. A general result claims for the importance of perceived compatibility between elderly persons' needs and environmental characteristics; when lacking, the consequence is often a dramatic decrease in the restorative (and healthy) potential of everyday settings. This can be true especially for the older elderly and, in general, impaired people.

A shortcoming of the present study is that the main focus was on the young elderly with no remarkable impairment. An interesting improvement would be to compare the restorative experiences of people at the "Third" vs. "Fourth" age (Baltes & Smith, 1999). Another aspect to be considered is that the sample mainly consisted of people living with spouse, and the role of social interaction in restorative experiences may presumably be somewhat different when people live alone. In addition, the task requires a reconstruction, which undoubtedly may imply a distortion from real experiences, and different findings might emerge in a field study. A final aspect is that we considered people from a large town, and different results might be outlined from a different residential and cultural context (e.g., a small town), which undoubtedly may affect everyday experiences.

In spite of some methodological limitations, the study addresses important dimensions of elderly persons-environment transactions. In addition, it identifies environmental features (natural elements, arts, etc.) which can be usefully included in guidelines for designing more healthy environments and suggests the importance of a place-specific regulation of social interaction in order to enhance positive outcomes. However, in order to plan and develop interventions in real settings, these findings have to be integrated with a more contextual analysis of residential areas needing to be improved, and specific demands of elderly people living there.

REFERENCES

Allard, J., Allaire, D., Leclerc, G. & Langlois, S. (1995). The influence of family and social relationships on the consumption of psychotropic drugs by the elderly. *Archives of Gerontology and Geriatrics, 20*, 193-204.

Avlund, K., Lund, R., Holstein, B. E. & Due, P. (2004). Social relations as determinants of onset of disability in aging. *Archives of Gerontology and Geriatrics, 38*, 85-99.

Baltes, P. B., & Smith, J. (1999). Multilevel and systemic analyses of old age: Theoretical and empirical evidence for a fourth age. In Bengtson, V. L. & Schaie, K. W. (Eds.), *Handbook of theories of aging* (pp. 153-173). New York: Springer.

Canter, D. (1986). Putting situations in their place: Foundations for a bridge between social and environmental psychology. In Furnham, A. (Ed.), *Social behavior in context* (pp. 208-239). Boston: Allyn and Bacon.

Christensen, D. L. & Carp, F. M. (1987). PEQI-based environmental predictors of the residential satisfaction of older women. *Journal of Environmental Psychology, 7*, 45-64.

Evans, G. W., Kantrowitz, E. & Eshelman, P. (2002). Housing quality and psychological well-being among the elderly population. *Journal of Gerontology: Psychological Sciences, 57B*(4), P381-P383.

Gitlin, L. N. (2003). Conducting research on home environments: Lessons learned and new directions. *The Gerontologist, 43*(5), 628-637.

Hammitt, W. E. (1984). Functions of privacy in wilderness environments. *Leisure Sciences, 6*(2), 151-166.

Hammitt, W. E. & Madden, M. A. (1989). Cognitive dimensions of wilderness privacy: A field test and further explanation. *Leisure Sciences, 11*(4), 293-301.

Hartig, T., Johansson, G. & Kylin, C. (2003). Residence in the social ecology of stress and restoration. *Journal of Social Issues, 59*(3), 611-636.

Hartig, T., Korpela, K., Evans, G. W. & Gärling, T. (1997). A measure of restorative quality in environments. *Scandinavian Housing & Planning Research, 14*, 175-194.

Herzog, T. R. (1985). A cognitive analysis of preference for waterscapes. *Journal of Environmental Psychology, 5*, 225-241.

Herzog, T. R., Chen, H. C. & Primeau, J. S. (2002). Perception of the restorative potential of natural and other settings. *Journal of Environmental Psychology*, 22, 295-306.

Herzog, T. R., Maguire, C. P, & Nebel, M. B. (2003). Assessing the restorative components of environment. *Journal of Environmental Psychology*, 23, 159-170.

Holmén, K. & Furukawa, H. (2002). Loneliness, health and social network among elderly people–a follow-up study. *Archives of Gerontology and Geriatrics*, 35, 261-274.

Horne, P., Boxall, P. C. & Adamowicz, W. L. (2005). Multiple-use management of forest recreation sites: A spatially explicit choice experiment. *Forest Ecology and Management*, 207, 189-199.

Kahana, E., Lovegreen, L., Kahana, B. & Kahana, M. (2003). Person, environment, and person-environment fit as influences on residential satisfaction of elders. *Environment and Behavior*, 35(3), 434-453.

Kaplan, R. & Kaplan, S. (1989). *The experience of nature: A psychological perspective*. New York: Cambridge University Press.

Kaplan, S. (1995). The restorative benefit of nature: Toward an integrative framework. *Journal of Environmental Psychology*, 15, 169-182.

Kaplan, S., Bardwell, L. V. & Slakter, D. B. (1993). The museum as a restorative environment. *Environment and Behavior*, 25(6), 725-742.

Kaplan, S. & Kaplan, R. (2003). Health, supportive environments, and the reasonable person model. *American Journal of Public Health*, 93(9), 1484-1489.

Kaplan, S. & Talbot, J. F. (1983). Psychological benefits of a wilderness experience. In I. Altman & J. F. Wohlwill (Eds.), *Behaviour and the natural environment* (pp. 163-203). New York: Plenum Press.

Korpela, K. M. (2002). Children's environment. In Bechtel, R. B. & Churchman, A. (eds.), *Handbook of Environmental Psychology* (pp. 363-373). New York: John Wiley & Sons, Inc.

Korpela, K. M. & Hartig, T. (1996). Restorative qualities of favorite places. *Journal of Environmental Psychology*, 16, 221-233.

Korpela, K., Kytta, M. & Hartig, T. (2002). Restorative experiences, self-regulation and children's place preferences. *Journal of Environmental Psychology*, 22, 387-398.

Kweon, B. S., Sullivan, W. C. & Wiley, A. R. (1998). Green common spaces and the social integration of inner-city older adults. *Environment and Behavior*, 30(6), 832-858.

Krause, N. (1996). Neighborhood deterioration and self-related health in later life. *Psychology and Aging*, 11(2), 342-352.

Laumann, K., Gärling, T. & Stormark, K. M. (2001). Rating scale measures of restorative components of environments. *Journal of Environmental Psychology*, 21, 31-44.

Oswald, F. & Wahl, H.-W. (2004). Housing and health in later life. *Reviews of Environmental Health*, 19(3-4), 223-252.

Perez, F. R., Fernandez, G. F-M., Rivera, E. P. & Abuin, J. M. R. (2001). Ageing in place: Predictors of the residential satisfaction of elderly. *Social Indicators Research*, 54(2), 173-208.

Purcell, T., Peron, E. & Berto, R. (2001). Why do preferences differ between scene types? *Environment and Behavior*, 33(1), 93-106.

Russell, J. A. & Ward, L. M. (1982). Environmental psychology. *Annual Review of Psychology, 33,* 651-688.

Scopelliti, M. & Giuliani, M. V. (2004). Choosing restorative environments across the lifespan: A matter of place experience. *Journal of Environmental Psychology, 24,* 423-437.

Scott, M. J. & Canter, D. V. (1997). Picture or place? A multiple sorting study of landscape. *Journal of Environmental Psychology, 17,* 263-281.

Staats, H. & Hartig, T. (2004). Alone or with a friend: A social context for psychological restoration and environmental preferences. *Journal of Environmental Psychology, 24,* 199-211.

Unger, J. B., McAvay, G., Bruce, M. L., Berkman, L. & Seeman, T. (1999). Variation in the impact of social network characteristics on physical functioning in elderly people: MacArthur studies of successful aging. *Journal of Gerontology: Social Sciences, 54B*(5), S245-S251.

Van den Berg, A. E., Koole, S. L. & van der Wulp, N. Y. (2003). Environmental preference and restoration: (How) are they related? *Journal of Environmental Psychology, 23,* 135-146.

Wallenius, M. (1999). Personal projects in everyday places: Perceived supportiveness of the environment and psychological well-being. *Journal of Environmental Psychology, 19,* 131-143.

Measures of Restoration in Geriatric Care Residences: The Influence of Nature on Elderly People's Power of Concentration, Blood Pressure and Pulse Rate

Johan Ottosson
Patrik Grahn

SUMMARY. In this paper, we have studied how elderly people's frame of mind influences their response to experiencing nature, measured in terms of blood pressure, pulse rate, concentration and results from protocols. Two theories concerning the importance of psychological balance have been put forward earlier by Lawton and Küller, both of whom maintain that the surrounding everyday environment is one of the keys to a harmonious existence. We present findings supporting the theory

Johan Ottosson, Agr Lic, landscape architecture, M Sc Hort, is Doctoral Candidate, Health & Recreation, Department of Landscape Planning, Swedish University of Agricultural Sciences.

Patrik Grahn, Agr.D, M.Sc (Biology), is Professor in landscape architecture, Health & Recreation, Department of Landscape Planning, Swedish University of Agricultural Sciences.

Address correspondence to: Johan Ottosson, Health & Recreation, Department of Landscape Planning, PO Box 58, Swedish University of Agricultural Sciences, S-230 53 Alnarp, Sweden (E-mail: Johan.Ottosson@movium.slu.se).

[Haworth co-indexing entry note]: "Measures of Restoration in Geriatric Care Residences: The Influence of Nature on Elderly People's Power of Concentration, Blood Pressure and Pulse Rate." Ottosson, Johan, and Patrik Grahn. Co-published simultaneously in *Journal of Housing for the Elderly* (The Haworth Press, Inc.) Vol. 19, No. 3/4, 2005, pp. 227-256; and: *The Role of the Outdoors in Residential Environments for Aging* (ed: Susan Rodiek, and Benyamin Schwarz) The Haworth Press, Inc., 2005, pp. 227-256. Single or multiple copies of this article are available for a fee from The Haworth Document Delivery Service [1-800-HAWORTH, 9:00 a.m. - 5:00 p.m. (EST). E-mail address: docdelivery@haworthpress.com].

Available online at http://www.haworthpress.com/web/JHE
© 2005 by The Haworth Press, Inc. All rights reserved.
doi:10.1300/J081v19n03_12

that the positive experience of natural surroundings per se has a balancing and healing effect.

We have found that the people most affected by their surroundings are those with the greatest psycho-physiological imbalance. When the balance tilts, the balancing effect of the green experience can restore the individual to a state of better harmony. Time spent in the outdoors is, thus, especially important for individuals who easily lose their equilibrium or find it difficult to make compensatory changes to restore harmony on their own.

The research project is an intervention study in which fifteen elderly individuals living at a home for very old people participated. Their power of concentration, blood pressure and heart rate were measured before and after an hour of rest in a garden and in an indoor setting, respectively. Seven elderly people were randomly chosen to have their first series of tests in a garden, while eight elderly people had their first series of tests indoors. The results indicate that power of concentration increases for very old people after a visit in a garden outside the geriatric home in which they live, as compared to resting indoors in their favorite room.

The results did not show any effects on blood pressure or heart rate.

However, when we compared these results with the background variables, we found interesting correlations. Background variables included how often they took part in social activities, showed helpfulness toward other residents in group activities and tolerance/critique of other residents. After a factor analysis, the background variables formed three distinct factors, one of which showed significant correlation with blood pressure and heart rate: the factor *psycho-physiological balance,* with the variables "degree of tolerance," "degree of helpfulness" and "frequency of hospital visits." Elderly people with low psycho-physiological balance, that is, who had low tolerance of other residents, were not helpful in group activities and had a high frequency of hospital visits, were most affected by a stay in a garden, as could be measured in changes in heart rate and blood pressure. The results may be interpreted as showing that a garden can restore an elderly person with low psycho-physiological balance to a state of better harmony.

The present study show, first, that an outdoor visit is important for recovery from stress and fatigue and, second, that the improvement is especially significant for the most susceptible. Thus, it is of particular importance that weak groups, such as elderly in great need of care, have access to an outdoor space. Such groups are likely to include many people who are in psycho-physiological imbalance. *[Article copies available for a fee from The Haworth Document Delivery Service: 1-800-HAWORTH. E-mail address: <docdelivery@haworthpress.com> Website: <http://www.HaworthPress. com> © 2005 by The Haworth Press, Inc. All rights reserved.]*

KEYWORDS. Restorative effects, garden, elderly people

INTRODUCTION

Ancient narratives throughout the world depict the garden and pastoral landscapes as places where you can take refuge in order to find shelter and relief from sadness and pain–places where you can be restored both mentally and physically. These narratives portray gardens and pastoral landscapes as healthy, healing places one longs for both in life and beyond (Stigsdotter & Grahn, 2002; Gunnarsson, 1992; Prest, 1988). In line with these traditional narratives, gardeners, occupational therapists, physiotherapists and others working in hospitals and clinics in the U.S., England, Japan and Sweden have been working with "horticultural therapy" and "nature therapy" (Grahn & Stigsdotter, 2003; Söderback et al., 2004). In the mid-1980s, some interesting research findings were published in the U.S.: It appeared that gardens, parks and areas of natural greenery did indeed have beneficial effects on people's health (Kaplan & Kaplan, 1989; Ulrich et al., 1991). The researchers called these effects "restorative."

Of special interest is one statement put forward by Roger Ulrich: "Persons who undergo medical treatment often feel psychologically vulnerable, which has been demonstrated to heighten their sensitivity to insecurity in an environment." And: "It seems likely that the restorative benefits of viewing nature are greatest when persons experience high levels of stress, such as those who are obliged to spend time confined in hospitals or other types of healthcare facilities" (Ulrich, 1999). Moreover, studies have shown that tolerance and helpfulness tend to decrease when people are under stress (Kaplan & Kaplan, 1989) or are mentally exhausted (Kaplan, 1990).

According to Ottosson (2001), who describes a long period of convalescence in a hospital after a traffic accident, visits to the out-of-doors played a significant role in recovery from fatigue and crisis in his rehabilitation after a brain injury.

Searles (1960) points out that signals from Nature spark creative processes that are important in the rehabilitation process. Complicated relations may be too much to handle. Most complex are our relations to other people, and the simplest relations are those between inanimate objects, such as stones, and us. Plants and animals fall somewhere in between. According to Searles, being able to master these relationships helps us to recover from crises (Andersson & Olsson, 1982; Searles, 1960).

This led us to formulate three hypotheses:

1. Being in the outdoors affects different people to different degrees.
2. The impact and significance of being in the outdoors will vary, depending on the individual's life situation.
3. An individual's preferences for features of the outdoors (solitude or being in a group; a sunny summer day or a violent autumn storm, etc.) will vary according to his/her frame of mind, that is, his/her capacity to absorb and process the impulses the experience involves.

Our aim was to study how a large and growing vulnerable group, elderly people in need of considerable care and attention, reacts to time spent in a natural environment.

EARLIER STUDIES

Some research findings suggest that ailing individuals and elderly people are more dependent than are others on certain characteristics of the physical environment in their neighborhood. Both Küller (1991) and Searles (1960) have demonstrated the importance of a familiar environment for the unwell and aged. Being able to continue to use his/her own furniture and familiar objects after moving to an institution heightens the individual's sense of well-being. Diary studies reported by Grahn (1989) demonstrate that the need for familiar surroundings applies to the outdoor environment as well.

Clearly, the resources at an individual's disposal influence his/her prospects of enjoying a rich and varied leisure time. The preconditions–the resource budget–consist of five different components (Grahn, 1991)–(See Figure 1).

One of the components is of special interest in this context:

• *Mental energy/constitution.* If an individual is under psychological stress, it may be difficult to achieve the composure necessary to pursue a rich variety of leisure activities. For example, the person may feel forced to live in the old age home against his/her will. This fifth component is often disregarded in assessments of people's prospects for a rich leisure, but for many it may be the most limiting factor.

FIGURE 1. The Budget of Resources that Influence the Content and Quality of One's Leisure (from Grahn, 1991)

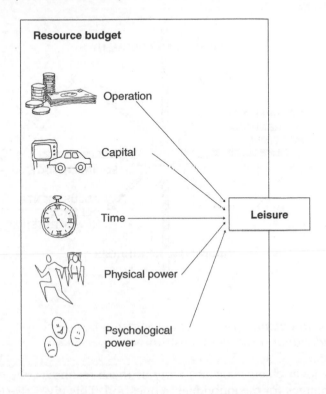

Küller (1991) points out three resources that should be taken into account in the planning of outdoor environments. They are medical, social and psychological factors, which together determine the individual's satisfaction with the environment. Küller (1991) has constructed a model showing how the resources interact (see Figure 2).

The interaction between the resources differs according to the person's age. The balance between activation and control varies over time. In order to maintain a feeling of harmony, the balance should not deviate too greatly from the person's needs for any considerable period of time.

The model includes four kinds of resources:

- *the physical environment* (home, neighborhood, community)
- *occupational & recreational activities* (household, leisure)

FIGURE 2. Rikard Küller's (1991) Model of Human-Environment Interaction, Comprising the Variables Activity, Individual, Social Surroundings and Physical Setting

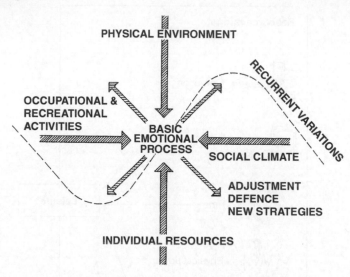

- *social climate* (partner, children, friends, neighbors)
- *individual resources* (constitution, experience)

Whenever changes occur in any of these resources, the overall balance changes for the individual in question. This gives rise to a need to change one or more of the other resources in order to restore balance. Change needs to be compensated for. The older the person, the more difficult it may be to make compensatory changes in response to unwelcome changes in his/her situation.

Lawton (1985) points out the importance of balance between familiar and new features in the environment. Whenever the unknown or the familiar is too dominant, we try to right the balance. According to Lawton, elderly people are more sensitive to this balance than are younger people. When an individual's flexibility (i.e., capacity to accommodate) declines, he/she has a greater need for compatible surroundings. Flexibility declines with increasing age, albeit with considerable individual variation.

A theory put forward by Tornstam (1997, 1998) treats older people's relationship with their environment, including the outdoor environment. Tornstam emphasizes the importance of what he calls "consensual validation"; that is, whereas older people have a definite idea of their identity, they require confirmation–validation–from features in their surroundings. When the older person's self-image is confirmed by signals from the environment, consensus is established. Older people have a greater need for familiar surroundings; their sense of security is more dependent on such familiarity–not least as regards the outdoor environment.

The experiences the Kaplans (1989) report from The Wilderness Laboratory show that mentally exhausted individuals are strongly influenced by their physical surroundings. Contact with untouched nature proved to be able to restore their mental equilibrium. The loss of mental equilibrium will result in exhaustion, what Kaplan (1990) calls "mental fatigue," which is characterized by, e.g., concentration difficulties, increased irritability and decreased likelihood of helping someone in distress.

Roger Ulrich also talks about the restorative power of the environment (Ulrich, 1983). He argues that the visual impact of the environment itself may signal danger or safety. An article he published in *Science* indicates that the view from a hospital over nature and green open spaces has a positive influence on recovery after surgery (Ulrich, 1984). His supporting findings (Ulrich et al., 1991; Ulrich, 1993, 1999, 2001) show that the body reacts spontaneously, within fractions of a second, to natural elements, whereas artifacts such as houses, streets, etc., do not trigger the same quick and strong reactions.

The Kaplans state that the restorative environment stimulates people's cognitive system, while Roger Ulrich discusses how effects on the physiological system are mediated by emotions.

Figure 3 shows the kind of recreation individuals appreciate, given different degrees of mental energy/constitution (Grahn, 1991). The y-axis shows mental energy/constitution and the x-axis shows degree of sensitivity to characteristics in the environment. At the bottom of the pyramid, the individual wants to be left alone, to be able to think things over or simply to relax in solitude. He/she seeks out more natural environments. This involves an important process, reflection, and an active relation with the environment (Ottosson, 1997, 2001).

Given somewhat more mental energy, the individual's need for nature will be less indispensable and specific, and the social needs and skills will be more apparent. People have different preferences with regard to the outdoors, depending on how much mental energy they have or their state of mental equilibrium (Grahn, 1991; Ottosson, 2001).

In the present study, we have used methods to measure people's power of concentration, in line with the theories of the Kaplans (1989),

FIGURE 3. Participation in Activities, in Relation to the Individual's Mental Energy and Capacity

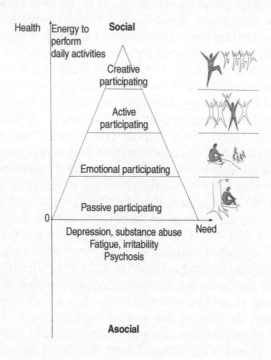

and we have measured blood pressure and pulse, in line with the theories of Ulrich (1993).

METHODS

In an intervention study of people living in an old age home (Ottosson & Grahn, 2005a), we found that residents' powers of concentration were quickly restored after a visit in the gardens surrounding the home. However, blood pressure and pulse rate did not show any significant change. According to the theory and laboratory findings of Ulrich (1999), we should have found lower blood pressure and pulse rate. We asked ourselves whether the background data on the old people might help explain the variation: These data consisted of questions posed to the residents and the staff, concerning preferences regarding their surroundings, the residents' past homes and how the elderly felt in general. On the basis of these data, the participants could be divided into subgroups.

The aim of the present intervention is to test our three hypotheses regarding elderly people in need of considerable care and attention, with regard to their sensitivity to their environment. We approach the task by comparing data on the elderly individuals' blood pressure, pulse rate and powers of concentration with background data gathered in interviews with the individuals themselves and with staff members at the old-age home.

We wished to focus on how very aged individuals, who require considerable assistance and care, respond to being in a "favorable outdoor environment," compared to resting indoors in their "favorite room."

Our overall study design is that of an intervention study. This kind of study commonly involves a separate intervention group and control group. One problem associated with this kind of study is that it would be difficult for us to find an intervention group and a control group that are identical with regard to age, gender, socioeconomic status, medication and other background factors, not least the factors focused on in this study: home and garden. The solution to this problem in our research design is to use the same individuals in the intervention group and the control group. Half of the participants started in the intervention group, that is, they rested in the outdoor setting. The other half started in the control group, that is, they rested in the indoor setting. After about a week, the study continued, however this time, the participants originally in the intervention group were now in the control group, and vice versa.

We needed an old-age home that was amenable to our research interests and whose residents were healthy enough to participate in the study. The physiological and psychological influence of the outdoor experience was tested by measuring blood pressure and heart rate, in line with Ulrich's theory, and via a battery of attention tests, in line with Kaplan and Kaplan's theory.

It was important to us that the participants find the study a pleasant diversion in their everyday lives. This was important from an ethical point of view, as well as necessary from a practical point of view, as we needed to pay an additional two visits. In our opinion, we succeeded in this effort.

This group of aged people in need of considerable care and attention was hard to get in contact with. One main reason is that their inclination to have strong routines in their everyday life is important (Berg et al., 1991). Breaking these routines by introducing new activities is difficult.

The Venue

We contacted the old-age homes located in an area known as Mårtenslund in the city of Lund in southern Sweden. The staff was interested in helping us. Both management and others in a position to help us found the study interesting.

The Mårtenslund complex consists of a cluster of four relatively independent old-age homes housed in separate buildings on a single block. In the complex, residents live in rooms of their own with their own personal furnishings. Figure 5 shows a floor plan of a residence in the Mårtenslund old-age home. Outside the room, a corridor links a number of rooms on the same floor. Some such rooms share spaces for gatherings, dining rooms and kitchen facilities.

All in all, the Mårtenslund area is home to several hundred elderly residents, all of whom require care–inasmuch as only those who cannot manage a household on their own are eligible to live there. Due to the infirmities of old age, it was at first difficult to find enough residents who were both willing and able (especially concerning visual handicap) to carry out our tests.

Although Mårtenslund is centrally situated in the city, the buildings are surrounded by an extensive park (Figure 4). Each building has a sheltered terrace near the entrance. The center of the area is a lawn with several old fruit trees, apple and pear. The lawn itself contains many flowers: daisies, buttercups and veronica. The side of the block facing a road has a tier of trees and bushes that shelters the block against traffic noise, which at times is quite heavy.

We gathered climatic data from the Swedish Meteorological and Hydrological Institute, as they have a measurement instrument located in the neighborhood in which we performed the study. The climatic data included temperature, light exposure and wind on the actual study days and hours when we were outdoors with the participants. On the days we studied the participants outdoors, there were no extreme weather conditions.

Composition of the Group

Seventeen of the interested residents participated in the study. One subject was unable to complete all the tests, and another participated at only one measurement. Both were eliminated from the study. Women predominated: 13 women and 2 men. This distribution reflects the ac-

FIGURE 4. Plan of the Block in Lund Where the Study Was Carried Out

FIGURE 5. Floor Plan of a Residence in the Mårtenslund Old-Age Home

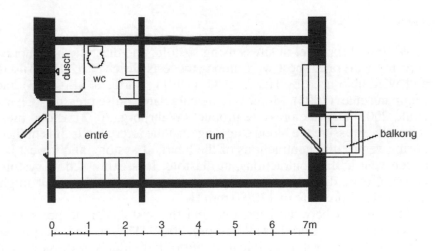

tual distribution among the residents of Mårtenslund (verbal information from the director of recreation, Mårtenslund). The age distribution is shown in Table 1.

The mean age is 86, and the median age 87. Four of the old people used wheelchairs, 11 did not.

TABLE 1. The Sex and Age of the Subjects, in Order of Age

Sex	M	F	F	F	F	F	F	F	F	F	M	F	F	F	F
Age	67	73	81	82	84	85	86	87	89	89	91	92	93	94	97

It was not easy to find "windows of opportunity" in our participants' daily schedules; this was due to, e.g., the participants' strong routines (Berg et al., 1991). As a consequence, the study took six months to carry out.

Background Data

Our participants were asked a set of questions that were posed in the form of a structured interview in which they responded orally. They were asked their age, sex and selected demographic characteristics. Finally, they were asked to rate the degree to which they felt at home in the building and its grounds. The ratings were made on a 7-point scale (Barnett, 1991); where 7 was "quite at home" and 1 "not at all." This scale functioned well.

Tests

We tested for level of stress using systolic and diastolic blood pressure and heart rate. These were among the tests Ulrich et al. (1991) used, and were also used by Hartig et al. (1991), Hartig et al. (2003) and Laumann et al. (2003). Moreover, we calculated pulse pressure (Chang et al., 2003) and rate pressure product (Währborg, 2002). Blood pressure is the pressure the blood exerts against the artery walls. It is highest during ventricular contractions of the heart, at systole, and lowest between ventricular contractions, at diastole. It is expressed as systolic pressure over diastolic pressure; for example, a healthy adult might have a blood pressure of 120/80 mm Hg.

The product between heart rate and the systolic blood pressure is called rate pressure product (Währborg, 2002). This product has proven to be reliable and valid in clinical studies of individuals in stressful situations (ibid). Pulse pressure is the difference between systolic pressure and diastolic pressure. The reliability and validity of these measures in old people are high (Chang et al., 2003).

These kinds of cardiovascular measures record activity that is controlled by the autonomic nervous system. This system is subdivided into the sympathetic and the parasympathetic nervous system. The major function of the sympathetic system is to mobilize the body for action, so that challenging or stressful situations can be dealt with efficiently. Sympathetic activation consumes energy and is, accordingly, physically taxing and non-restorative; it includes: increased heart rate, contraction of blood vessels in the skin and increased blood pressure, both systolic and diastolic (Ulrich et al., 1991).

In addition, we used the following measures of concentration, selected on the basis of their focus and demonstrated reliability and validity (Kuo, 1992): The Necker Cube Pattern Control Test (NCPC), Digit Span Forward (DSF), Digit Span Backward (DSB) and The Symbol Digit Modalities Test (SDMT) (Ottosson & Grahn, 2005a).

Questions to Staff Members

To find out whether differences in test results within the age group might be correlated with certain background variables, we divided the group into categories.

For the most part, the division into subgroups follows Küller's (1991) model including variables pertaining to "Occupational & recreational activities," "Social climate" and "Individual resources," respectively. In addition, we use Grahn's (1991) categorization of individuals' constitution in terms of "Physical condition" and "Mental energy." The variables used to fill out this framework were generated by us, and by staff members who both had extensive professional experience and were well acquainted with the aged participants. We took pains to ensure that the classification was made in a non-arbitrary manner. The form was structured so the variables would be easy to understand, check and fill out.

In time, we formed a list of variables that could be registered by indicating the presence or absence of the variable in question. Tolerance, for example, can be checked in terms of behaviors and events that occur/do not occur in conjunction with meals in the common dining room. Thus, the classification consists exclusively of notations of presence/absence of variables. The recreation director and the staff performed the classification.

The subgroups are shown in Table 2.

Procedure

The study proceeded as follows: With the help of the recreation director, we established contact with a number of residents who were interested in participating. The windows of opportunity for carrying out the tests were very narrow: more or less the time between the morning

TABLE 2. Background Variables Generated in Interviews with Staff

1. Variables pertaining to **Social climate**

1a *Children*, two categories:
Have no children = 0
Have children = 1

1b *Married or cohabiting during active years*, two categories:
Married = 1
Unmarried = 0

1c *Employed outside the home or at home during active years*, two categories:
Employed outside the home = 1
At home = 0

2. Variables pertaining to **Occupational & recreational activities**

2a *Territory covered in walks*, 3 categories:
Never outdoors = 0
Goes out, but stays close to home = 1
Takes long walks = 2

2b *Degree of participation in group activities at old-age home*, 4 categories:
Never takes part = 0
Seldom takes part = 1
Takes part fairly often = 2
Takes part regularly = 3

2c *Relations with family*, 2 categories:
Good relations = 1
Absence of good relations = 0

3. Variables pertaining to **The individual's mental energy/constitution**

3a *Lucidity/confusion*, 2 categories:
Lucid, rational = 1
Not lucid = 0

3b *Helpfulness toward other residents*, 2 categories:
Generally helps others = 1
Not usually helpful = 0

3c *Tolerance toward residents in group activities*, 2 categories:
Behaves tolerantly = 1
Does not behave tolerantly = 0

4. Variables pertaining to **The individual's physical condition/constitution**

4a *Degree of physical disability*, 4 categories:
Requires wheelchair = 0
Requires walker = 1
Requires cane = 2
Requires no aids = 3

4b *Eyesight*, 2 categories:
No problems or minor problems = 1
Disability that requires special aids = 0

4c *Frequency of visits to the hospital*, 3 categories:
Never goes to the hospital = 2
Occasional visits to the hospital – 1
Frequent visits to the hospital = 0

break and lunch. As a consequence, all the testing was done during mid-morning.

We tested our participants:

• Immediately before and after they had spent time in the garden.
• Immediately before and after they had spent time indoors.

The period of recreation between tests was roughly one hour. Each participant was tested on three different days. The first day was used to measure the effect of having spent time in the garden or indoors. The second test day was devoted to the corresponding test, indoors or out-doors. The third test day was used for the questionnaires. For each participant, all three test days occurred at intervals of approximately 14 days. In this design, the respondents served as their own controls.

Participants were to spend their hour of leisure in the environment they enjoyed the most, indoors or outdoors. During the leisure periods indoors and outdoors, the individuals tested were not allowed to take any medicine, drink coffee or sleep. Both indoors and outdoors, the individuals were sitting most of the time, resting in a chair or on a bench. The activity was limited to chatting with us.

On the third day we met, the day for questionnaires, we asked participants some background questions, using the questionnaires.

Finally, staff members and the director of recreation were asked to supply background data about the participants.

Figure 6 shows the tests we conducted. Seven elderly people accomplished the study by having their first series of tests *outdoors*, while

FIGURE 6. The Design of the Study, with Three Rounds of Tests

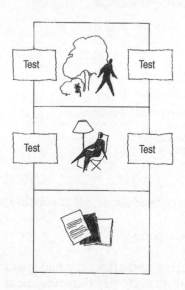

eight elderly people started their study *indoors*. Hence, approximately half of the group started indoors, half of the group outdoors. The time span between the two series of tests was 7-11 days.

Diastolic blood pressure, systolic blood pressure and heart rate were all measured in the upper left arm using a device of the type Biocomfort Automatic Control OM 5 (Biocomfort, 1995). It consists of an inflatable rubber collar fitted with a microphone, connected to a manometer. The system is computer-steered. The systolic and diastolic blood pressure and heart rate are registered and read to the computer memory automatically. Results from the measures above were used to calculate the participants' pulse pressure (Chang et al., 2003) and rate pressure product (Währborg, 2002).

The four concentration tests were: the Necker Cube Pattern Control Test, the Digit Span Forward Test, the Digit Span Backward Test and the Symbol Digit Modalities Test (Ottosson & Grahn, 2005a).

One factor, which may have influenced concentration, is greater circulation of blood through the body and brain due to differences in physical activity. Although the elderly participants did not move their bodies a great deal, those who are pushed in wheelchairs do not move their body more when visiting a garden compared to being indoors. To make an extra check of whether there could be any possibility of finding differences concerning body movement, we made a comparison of elderly who use wheelchairs and elderly who do not.

Data Analyses

We have had to take measures to accommodate the fact that the observations are few. Consequently, we have chosen to apply several non-parametric methods. The statistical significance of the differences between before and after values was calculated using the Wilcoxon Rank Sum Test (Ferguson, 1984). Values were fed into a raw data file, which was then processed using statistical data processing programs (SAS Statistics, 1996).

RESULTS

Categorization of the participants produced a set of background variables (referred to as *balance variables* in the following). For the sake of clarity, we sought to reduce the number of variables. Could they be clustered into meaningful groups? To find out we applied a factor analysis procedure (SAS Statistics, 1996). The variant of factor analysis used was based on principal components, and was orthogonal, varimax rotated, with the number of factors decided by eigenvalues exceeding one (Manly, 1994). The analysis reveals relationships between the variables. In the present case, the analysis produced three groups or factors.

The figure to the right in the table, whether positive or negative, indicates the explanatory value of the variable. Values under +/− 0.50 are considered weak indications and have been placed within parentheses here.

The most important variable in Table 3 is "married." Whether the individual has had a partner during his/her active years, whether the individual has children and the nature of his/her relations with family members constitute one aspect of the factor. The other, which is positively correlated, has to do with the area covered in the individual's walks outdoors and how often he/she takes part in group activities. The first factor seems to concern social contacts. This factor will be referred to in the following as "Social balance."

The dominant variable in the second factor in Table 4 is the individual's level of tolerance. This is followed by frequency of visits to the hospital and how helpful the individual is. Stephen Kaplan (1990)

TABLE 3. Factor 1. Social Balance. Staff Assessments of Background Variables. SAS Principal Component Analysis.

Factor 1	
Territory covered	0.72
Group activities	0.58
(Family relations	−0.46)
Children	−0.80
Married	−0.87

TABLE 4. Factor 2. Psycho-physiological balance. Staff assessment of background variables. SAS Principal Component Analysis.

Factor 2	
Tolerance	0.92
Hospital visits	0.76
Helpfulness	0.55
(Work	0.35)

writes about a mental fatigue that it is basically a question of low powers of concentration. Among other things, "They are irritable and less likely than usual to help someone in distress" (Kaplan, 1990). The members of this group are not particularly "stable." Level of tolerance, frequency of visits to the hospital and readiness to help others are the ingredients in this factor. We refer to this second factor as "Psycho-physiological balance."

The third factor, described in Table 5, contained variables representing degrees of functional disability. We call it "Physical balance."

By multiplying the component variables in the factors by their principal component values, we obtain three balance factors (Morrison, 1976).

Social Balance
M = married, C = children, SB = social balance
SB = (M × 0.868) + (C × 0.803)

Psycho-physiological Balance
T = tolerance, HV = hospital visits, H = helpfulness
PB = psycho-physiological balance
PB = (T × 0.920) + (HV × 0.764) + (H × 0.546)
Physical Balance
PH = physical handicap, PB = physical balance
PB = (PH × 0.630)

The next step was to see whether the different subgroups, defined in terms of the balance factors or questionnaire responses, performed differently on the concentration tests and measures of blood pressure and pulse rate, i.e., we were interested in seeing whether we could find any correlations.

The test results had been reduced to differential quotients, i.e., to the differences in performance after rest indoors and rest outdoors, respectively. Table 6 shows the correlations obtained between, e.g., Social balance and the Necker Cube test. A high value, whether positive or negative (+ or −0.50 to 1.00), indicates a strong association between the two variables. Low values (close to zero) indicate no mutual relation. The table also shows the p-values, indicating level of significance. "ns" stands for "not significant."

When we tested the three groups of factors against the measurements of power of concentration, blood pressure and pulse rate, we found four significant correlations: between psycho-physiological balance and pulse rate, diastolic blood pressure, pulse pressure and rate pressure product, respectively (see Table 7).

Thus, the analysis shows that diastolic blood pressure, pulse rate, pulse pressure and rate pressure product are influenced significantly by a period of rest in a garden in elderly people in need of considerable care and attention.

TABLE 5. Factor 3. Physical balance. Staff assessments of background variables. SAS Principal Components Analysis.

Factor 3	
Physical handicap	0.63
(Poor eyesight	−0.33)
(Lucidity	−0.43)

TABLE 6. Spearman Correlations Between Differential Concentration Scores and Balance Factors

	NCPC	DSF	DSB	SDMT
Social balance	0.06ns	0.07ns	−0.08ns	0.06ns
Psycho-physiological balance	−0.21ns	0.31ns	−0.16ns	−0.21ns
Physical balance	0.26ns	−0.31ns	−0.01ns	0.16ns

TABLE 7. Spearman Correlations Between Balance Factors and Differential Blood Pressure, Pulse Rate and Restorative Evaluation

	Systolic blood pressure	Diastolic blood pressure	Pulse	Pulse Pressure	Rate Pressure Product
Social balance	0.04ns	0.11ns	0.13ns	0.13ns	0.16ns
Psycho-physiological balance	0.04ns	0.52 $p < 0.05$	0.79 $p < 0.001$	0.68 $p < 0.002$	−0.46 $p < 0.02$
Physical balance	0.26ns	0.08ns	0.05ns	−0.17ns	−0.15ns

In Table 8 and Figures 7 and 8, we calculated the stability of the two basic values (diastolic blood pressure and pulse rate) by studying what happened when we excluded each participant, one at a time, from the analysis. We found that no one person exerted such an influence on the correlation between psycho-physiological balance and pulse rate that the significant result was altered. But six individuals did exert such an influence on the correlation between diastolic blood pressure and psycho-physiological balance. When any one of the six was excluded, the correlation was no longer significant. Thus, the values relating to pulse rate were surprisingly stable, considering the small number of participants.

We plotted the two basic values in a diagram (see Figures 9 and 10). Instead of column graphs, which are difficult to assess visually, we used so-called polygon lines. The top values in each column are linked together to form a curve. Two curves and a base-line represent each of the two variables diastolic pressure and pulse rate: one base-line for each of

TABLE 8. Calculation of Stability. The figure in the column headed Total correlation indicates the correlation value when each value has been eliminated from the material. For example: If one removes the first value, where the balance value is 0 and diastolic blood pressure is 7.5, the correlation would be 0.47.

Balance	Diastolic pressure	Total correlation	Pulse rate	Total correlation
0	−7.5	0.47	−2.5	0.78
0	3.5	0.62	−10.5	0.74
1.53	1.5	0.54	−1	0.78
1.53	3.5	0.60	0	0.78
1.68	−8	0.53	−6.5	0.78
2.45	7	0.53	6	0.83
2.45	2	0.53	−5.5	0.83
2.45	−5.5	0.54	4.5	0.81
2.45	−7	0.54	1	0.77
2.45	−28.5	0.58	0.5	0.77
2.99	8.5	0.44	3.5	0.80
2.99	2	0.56	2	0.80
2.99	6.5	0.46	7.5	0.76
2.99	9.5	0.44	5	0.78
2.99	6.5	0.46	18.5	0.76
N = 15		0.52		0.79

FIGURE 7. Diagram showing the stability calculation for diastolic blood pressure. The higher the value in the figure, the more certain it is. The top nine values are statistically significant.

Diastolic blood pressure: 9 values significant

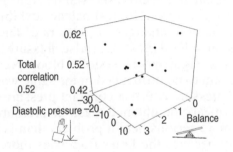

FIGURE 8. Diagram showing the stability calculation for pulse rate. All values are statistically significant.

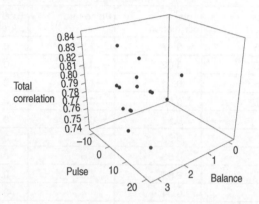

the two tests before the rest period, one curve for the values after a period of rest indoors, and one curve for the values after a period of rest in the garden.

We find that the poorer the psycho-physiological balance value, the less the individual benefited from rest indoors. Rest outdoors seems to have restored both pulse rate, diastolic blood pressure, pulse pressure and rate pressure product, whereas these values continued to rise during the period of rest indoors. Restoration for individuals with good psycho-physiological balance values was minor.

DISCUSSION

Social and physical balance factors showed no significant relationships with any test. Psycho-physiological balance exhibited a surprisingly strong correlation with changes in pulse rate. The correlations with changes in diastolic blood pressure, pulse pressure and rate pressure product were also significant. Systolic blood pressure, on the other hand, did not seem to be influenced in the same way in our sample. Diastolic blood pressure exerts a constant pressure on the blood vessels between heartbeats; heightened diastolic pressure is therefore considered a much more serious health problem than is high systolic pressure, particularly because the latter fluctuates more and is more

FIGURE 9. Diagram showing mean pulse rate, given different balance values at the time of the test. The 0-line indicates the mean value of pulse rate in the individual at rest before the tests.

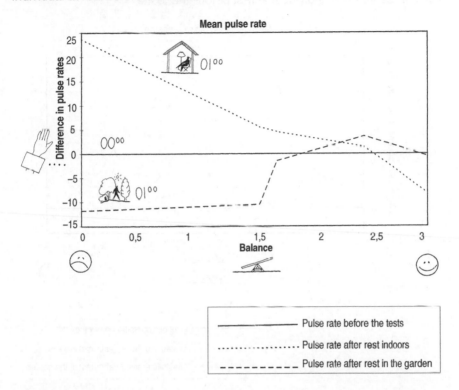

easily affected by external factors (Archives of Internal Medicine, 1984).

When we plotted the values recorded for diastolic blood pressure and pulse rate against the psycho-physiological factor, we found that pulse and blood pressure were quickly restored when the aged participants rested in the garden. When they rested indoors, however, the values continued to rise. These effects were stronger among those who had a poorer psycho-physiological balance at the outset.

Pulse rate and blood pressure increase in response to stress. The tests we put the participants through were probably stressful. Rest in the garden restored participants' values, i.e., their bodies coped with the stress. Given rest indoors, the stress persisted among those judged to have poor

FIGURE 10. Mean values of diastolic blood pressure, given different balance values at the time of the test. The 0-line indicates the mean value of diastolic blood pressure in the individual at rest before the tests.

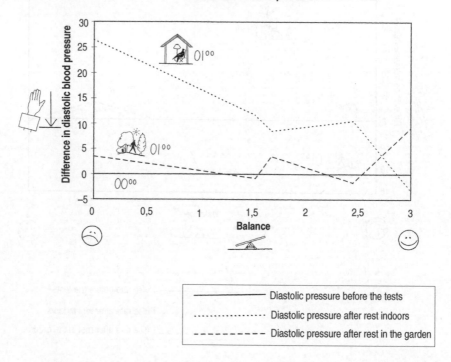

psycho-physiological balance. In this case, the values continued to rise to a significantly higher level one hour later.

The following studies support our results:

* Findings concerning the restorative power of nature experiences may be compared with results from a laboratory experiment reported by Ulrich et al. (1991). In that experiment, test persons were subjected to a high level of stress. They were then allowed to watch and listen to nature documentaries or films set in an extremely urban environment. All the subjects physiological values were restored to normal after only four to seven minutes of watching the nature films, whereas the films in urban settings had little or no restorative effect–some values

continued to rise until the experiment was called off. Hartig et al. (2003) obtained similar findings from a field experiment.

- Küller and Küller (1994) looked for possible correlations between elderly people's health status and their contact with the outdoors. Elderly people who looked out the window with a view of greenery had significantly higher health index scores than did those with a view of non-greenery.
- Rodiek (2002) conducted a study of the influence of an outdoor garden on stress in old people at a residential complex for the elderly. She found that cortisol levels were significantly lower in garden settings compared to indoor settings. This may be an indication that frailer elderly individuals are more dependent on a green environment.
- A study by Rappe and Kivelä (2005) on elderly long-term care residents found that "affective effects of visiting the garden tended to be more pronounced among the depressed than among those not depressed" (ibid p. 298). Moreover, they found that "the depressed felt more balanced and more cheerful and alert after visiting the garden" (ibid p. 302).
- Ottosson and Grahn (2005b) conducted a questionnaire study in which 547 people completed a validated protocol developed to evaluate individuals who have survived a crisis, brought about in connection with trauma, illness or depression. The protocol comprised several parts, which indicate, e.g., the frequency of stress conditions and rehabilitation potential. The questionnaire also comprised questions concerning everyday activities, such as how often the person experiences nature and how often the person feels he/she is critical of others. Results show clearly and significantly that: the rehabilitation potential of those greatly affected by crises have the strongest relationship to: "Having many experiences of nature in everyday life." The results could be interpreted as follows: *experiencing nature* seems to have a more powerful influence on the rehabilitation potential of people greatly affected by a crisis. *Taking walks in nature* does not seem to be of equal importance. *The social factor* seems to have more influence on the rehabilitation potential of people affected by a crisis to a low/moderate degree.

Comparing the present results to those of Ottosson and Grahn (2005a), we may wonder why a period of rest outdoors improved all the aged participants' powers of concentration, whereas the positive effect on pulse rate, diastolic blood pressure, pulse pressure and rate pressure product was more pronounced among those with relatively poor psycho-physiological balance.

It may be that several systems of nerve paths, brain functions and perceptual faculties are involved. They may sometimes operate in concert, and sometimes give rise to different kinds of reactions. The systems we are particularly interested in have to do with feelings, emotions and reflex reactions.

Emotions–and the behaviors they give rise to–display patterns other than logical thinking and the behaviors it gives rise to: This has been recognized ever since antiquity. For this reason, Descartes sought to draw a definite line between logic/rationality and emotion in his thinking about knowledge and the scientific method (Fredriksson, 1994). Goleman (1995) and Gardner (1993) consider emotions essential parts of the human intellect. Coming to terms with our emotions is a means of developing the intellect. Emotions help us to orient ourselves in the world around us and to communicate, both with others and with ourselves, through a so-called "inner dialogue."

The description of Ottosson (1997, 2001) is one of strong feelings about the relationship with different parts of Nature in a life crisis. We receive signals from Nature that are very important and direct, even though we may not consciously perceive them. Searles (1960) points out that people in crisis need stable environments in order to feel well. In situations of crisis, the individual may need to revert to simpler relations in Nature. Searles' conception of Nature as a link between the conscious and the subconscious is of special relevance in this context: According to Searles (1960), contact with Nature can contribute substantially to people's recovery from critical situations of various kinds. It sparks creative processes that are important in the rehabilitation process. This, says Searles, helps to reduce anxiety and pain, restore the sense of self, improve our perceptions of reality and promote tolerance and understanding (Andersson & Olsson, 1982).

In physiological terms, emotions are largely located in the older part of the brain, in the limbic system, which is the product of millions of years of evolution. Directly adjacent to the limbic system is an even older system, the brainstem (Bergström, 1992; Hjort, 1993). Directed attention (Kaplan, 1990), the mental process we use to deal with cognitive data, originates in a more "modern" part of the brain. Its effect on the autonomic nervous system and endocrine glands is less immediate. Feelings, particularly primitive impulses such as the impulse to flee or to seek food, have been important to the survival of the human race. These feelings originate for the most part in the oldest parts of the brain.

They give rise to very quick reactions in the nervous system, which in turn stimulate the autonomic nervous system and endocrine glands (Bergström, 1992; Hansson, 1996).

A number of researchers have related our emotions and feelings to more automatic "reflex" behaviors–the impulse to flee, stress, feelings of insecurity, the sense of danger and "home feeling," etc.–to conceptions of Homo sapiens' first habitats (Coss & Moore, 1990; Coss, 1991; Ulrich, 1993; Kaplan & Kaplan, 1989; Sorte, 1995; Hjort, 1983). According to many of these scholars, our primeval habitat was on a savannah, in proximity to water (ibid). The oasis, they propose, was surrounded by grasslands, punctuated by large, old trees. "Home" was located in protective green surroundings and commanded a view of water and the surrounding terrain–predominantly lightly forested, open fields. Many modern parks and gardens–and not least Swedish pastureland–fit that description.

Our theory is that individuals who are not in psycho-physiological balance are more prone to sense danger and to feel the impulse to flee. Consequently, they subconsciously seek security in the primeval home– a place, perhaps a garden, with light, open spaces such as meadows or pasturelands. Thus, these individuals might be expected to respond, in terms of blood pressure and pulse, more strongly to time spent in such a setting. Restoration of powers of concentration, directed attention capacity, on the other hand, involves the cognitive capacity to process information. Thus, all members of the group are affected equally (Ottosson & Grahn, 2005a).

It is conceivable that a positive experience of natural surroundings in itself has a balancing or buffering effect. Regardless of the direction in which the balance tilts, the experience can restore the individual to a better state of harmony. Time spent in the outdoors is, thus, especially important for individuals with low psycho-physiological balance, individuals who easily lose their equilibrium or who find it difficult to make compensatory changes to restore harmony on their own.

ACKNOWLEDGEMENTS

The authors would like to thank a number of people who have helped make the study and the report possible: Director of Recreation at the Mårtenslund home, Thomas Göransson, who served as their contact person; Associate Professor in statistics, Jan-Eric Englund, who has advised them concerning the statistical treatment of the material. The authors would also like to express their sincere thanks to the respondents who answered the questionnaire and the residents who participated in the study.

The research reported here was supported by the Swedish Research Council for Environment, Agricultural Sciences and Spatial Planning, Grant No. 2001-0252 and the Department of Landscape Planning at Alnarp, and SLU Kontakt.

REFERENCES

Andersson, T. and Olsson, E. 1982. *Äga rum.* Examensarbete, Stockholm: Psykologiska institutionen, Stockholms universitet.

Archives of Internal Medicine. 1984. Report of the Joint National Committee on the Detection of the Evaluation and the Treatment of High Blood Pressure. *Archives of Internal Medicine* 144: 1045.

Barnett, V. 1991. *Sample Survey Principles and Methods.* New York: Oxford University Press.

Berg, S., Carlsson, M. and Wenestam, C-G. 1991. The Oldest Old: Patterns of Adjustment and life Experiences. *Scandinavian Journal of Caring Science* 5: 203-210.

Bergström, M. 1992. *Hjärnans resurser.* Jönköping: Brain Books.

Biocomfort. 1995. *Biocomfort Automatic Control OM 5.* Biocomfort Blutdruck-Computer BF 80-OM 5 serienr B 501299, Gesundheitspflege GmbH, Hauptstrasse 18, 7300 Esslingen, Germany.

Chang, J.J., Luchsinger, J.A. and Shea, S. 2003. Antihypertensive medication class and pulse pressure in the elderly: analysis based on the third National Health and Nutrition Examination Survey. *American Journal of Medicine* 115: 536-542.

Coss, R. G. 1991. Evolutionary Persistence of Memory-Like Processes. *Concepts in Neuroscience* 2: 129-168.

Coss, R.G. and Moore, M. 1990. All that glistens: Water connotations in surface finishes. *Ecological Psychology* 2: 367-380.

Engfeldt, P. and Östergren, J. 1997/1998. Hypertoni. In: *Läkemedelsboken* 276-286. Stockholm: Apoteksbolaget.

Ferguson, G.A. 1984. *Statistical Analysis in Psychology and Education.* Singapore: McGraw-Hill International Book Company.

Fredriksson, G. 1994. *20 filosofer.* Stockholm: Pan Norstedts.

Gardner, H. 1993. *Multiple Intelligences.* New York: Basic Books.

Goleman, D. 1995. *Känslans intelligens.* Stockholm: Wahlström & Widstrand.

Grahn, P. 1989. *Att uppleva parken.* SLU, Institutionen för landskapsplanering, 89:6, SLU, Alnarp.

Grahn, P. 1991. *Om parkers betydelse.* Dissertation. Stad & Land 93, Movium, Alnarp.

Grahn, P and Stigsdotter, U. 2003. Landscape Planning and Stress. *Urban Forestry & Urban Greening* 2, 1-18.

Gunnarsson, A. 1992. *Fruktträden och paradiset.* (Dissertation.) Stad & Land, Movium/instttutionen för landskapsplanering, Sveriges Lantbruksuniversitet, Alnarp.

Hansson, L.Å. 1996. Psykoneuroimmunologi. Örebro Läkardagar om psykosomatik. Svensk Medicin no 52. SPRI, Stockholm.

Hartig, T., Mang, M., and Evans, G. 1991. Restorative Effects of Natural Environmental Experiences. *Environment and Behavior, 23*: 3-26.

Hartig, T. Evans, G.W., Jamner, L.D., Davis, D.S. and Gärling, T. 2003. Tracking restoration in natural and urban field settings. *Journal of Environmental Psychology, 23*, pp 109-123.

Hjort, B., 1983. Var hör människan hemma? Kungliga Tekniska Högskolan, Formlära, Stockholm.

Kaplan, R., Kaplan, S. 1989. The Experience of Nature. Cambridge University Press, Cambridge.

Kaplan, S. 1990. Parks for the future–A psychologist view. In Sorte, G. J. (Editor), Parks for the future. Stad & Land 85, Movium, Alnarp, pp. 4-22.

Kaplan, S. (2001) Meditation, restoration, and the management of mental fatigue. *Environment and Behavior, 33*, pp 480-506.

Kaplan, S. and Kaplan, R. 1982. Environment and Cognition. New York: Praeger.

Kuo, F. 1992. A Guide to Restorative Assessment Methodology. The University of Michigan, Ann Arbor.

Küller, R. 1991. Environmental assessment from a neuropsychological perspective. In: Gärling, T. and Evans, G.W. (Editors), *Environment, cognition and action: An integrated approach*. Oxford University Press, New York, pp. 111-147.

Küller, R., Küller, M. 1994. Stadens grönska, äldres utevistelse och hälsa. Byggforskningsrådet, R24:1994. Svensk Byggtjänst, Stockholm.

Laumann, K., Gärling, T. and Stormark, K.M. 2003. Selective attention and heart rate responses to natural and urban environments. *Journal of Environmental Psychology, 23*, pp 125-134.

Lawton, M.P. 1985. The Elderly in Context: Perspectives from Environmental Psychology and Gerontology. *Environment and Behavior, 17*: 501-519.

Lewis, C.A. 1996. Green Nature. Human Nature. Urbana, Chicago.

Manly, B.F.J. 1994. Multivariate Statistical Methods. A primer. Second edition. Chapman & Hall, London.

Mooney, P., and Nicell, P.L. 1992. The importance of exterior environment for Alzheimer residents: Effective care and risk management. Healthcare Management Forum, 5: 23-29.

Morrison, D.F. 1976. Multivariate Statistical Methods. Second edition. McGraw-Hill, New York.

Ottosson, J. 1997. Naturens betydelse i en livskris. Stad & Land 148. Movium, Alnarp.

Ottosson, J. 2001. The Importance of Nature in Coping with a Crisis. Landscape Research, 26: 165-172.

Ottosson, J. and Grahn, P. 2005a. A Comparison of Leisure Time Spent in a Garden with Leisure Time Spent Indoors: On Measures of Restoration in Residents in Geriatric Care. *Landscape Research*, 30: 23-55.

Ottosson, J. and Grahn, P. 2005b. Measures of Restoration in People Suffering from Crises: How does frame of mind influence the response to experiences of nature with regard to measures of restoration? Submitted.

Rappe, E. and Kivelä, S-L. 2005. Effects of garden visits on long-term care residents as related to depression. *HortTechnology* 15: 298-303.

Prest, J. 1988. *The Garden of Eden; the botanic garden and the re-creation of paradise*, Yale Univ. Press, New Haven.

Rodiek, S. 2002. Influence of an outdoor garden on mood and stress in older persons. *Journal of Therapeutic Horticulture*. 13: 13-21.

SAS Statistics. 1996. Version 6 Release 6.09. SAS Institute Inc., Cary, North Carolina.

Searles, H. F. 1960. The Nonhuman Environment in Normal Development and in Schizophrenia. International Universities Press, New York.

Simson, S.P., Straus, M.C. 1998. Horticulture as Therapy. New York, London.

Söderback, I., Söderström, M. and Schälander, E. 2004. Horticultural therapy: The 'healing garden' and gardening in rehabilitation measures at Danderyd Hospital Rehabilitation Clinic, Sweden. *Pediatric Rehabilitation*. 7 (4): 245-260.

Sorte, G.J. 1995. The value of nature and green spaces to the urban resident. Homo urbaniensis. In: *Ecological aspects of green areas in urban environments*. IFPRA world congress 3-8 September 1995, Antwerp, pp. 5.43-5.46.

Stewart, S. 2003. Horticultural therapy for patients with eating disorders at the Homewood Health Centre, Guelph, Ontario. *Journal of Therapeutic Horticulture*, 14: 32-37.

Stigsdotter, U. and Grahn, P. 2002. What Makes a Garden a Healing Garden? *Journal of Therapeutic Horticulture*, 13: 60-69.

Stigsdotter, U. A. and Grahn, P. 2003. Experiencing a Garden: A Healing Garden for People Suffering from Burnout Diseases. *Journal of Therapeutic Horticulture* 14: 38-49.

Tornstam, L. 1997. Life Crises and Gerotranscendence. *Journal of Aging and Identity*, 2: 117-131.

Tornstam, L. 1998. Åldrandets socialpsykologi. Rabén & Sjögren, Stockholm.

Ulrich, R.S. 1983. Aesthetic and affective responses to natural environments. In: Altman, I. and Wohlwill, J.F. (Editors), *Human Behavior and Environment*, Vol 6, New York, pp. 85-125.

Ulrich, R. S. 1984. View Through a Window May Influence Recovery from Surgery. *Science, 224*: 420-421.

Ulrich, R.S. 1993. Biophilia, biophobia and natural landscapes. In: Kellert, S.R. and Wilson, E.O., (Editors), *The Biophilia Hypothesis*, pp. 73-137.

Ulrich, R. S. 1999. Effects of Gardens on Health Outcomes: Theory and Research. In: Cooper Marcus, C. and Barnes, M., (Editors), *Healing Gardens. Therapeutic Benefits and Design Recommendations*, John Wiley & Sons, New York. pp. 27-86.

Ulrich, R. S. 2001. Effects of Healthcare Environmental Design on Medical Outcomes. In: Dilani, A. (Editor), Design & Health, Svensk Byggtjänst, Stockholm, pp 49-59.

Ulrich, R. S., Simons, R. F., Losito, B. D., Fiorito, E., Miles, M. A., and Zelson, M. 1991. Stress Recovery During Exposure to Natural and Urban Environments. *Journal of Environmental Psychology, 11*: 201-230.

Währborg P. 2002. Rate Pressure Product vid stress-en översikt. Konsensusmöte. *Institutet för Psykocosial Medicin och Arbetsmiljöverket 12-13 februari 2002.*

Index

Page numbers in *italics* designate figures; page numbers followed by the letter "t" designate tables.

© 2005 by The Haworth Press, Inc. All rights reserved.

T - #0498 - 101024 - C0 - 212/152/16 - PB - 9780789032447 - Gloss Lamination